Installer's Guide to Local Area Networks

Installer's Guide to Local Area Networks

THOMSON
DELMAR LEARNING™

Australia • Canada • Mexico • Singapore • Spain • United Kingdom • United States

Installer's Guide to Local Area Networks

Buddy Shipley

Vice President, Technology and Trades SBU:
Alar Elken

Editorial Director:
Sandy Clark

Senior Acquisitions Editor:
Gregory L. Clayton

Development:
Dawn Daugherty

Marketing Director:
Maura Theriault

Channel Manager:
Beth A. Lutz

Marketing Coordinator:
Brian McGrath

Production Director:
Mary Ellen Black

Production Manager:
Andrew Crouth

Production Editor:
Sharon Popson

Printed in Canada.
1 2 3 4 5 XX 05 04 03

For more information contact Delmar Learning
Executive Woods
5 Maxwell Drive, PO Box 8007,
Clifton Park, NY 12065-8007
Or find us on the World Wide Web at
http://www.delmarlearning.com

Library of Congress Cataloging-in-Publication Data:

ISBN 0-7668-3374-7

NOTICE TO THE READER

Publisher does not warrant or guarantee any of the products described herein or perform any independent analysis in connection with any of the product information contained herein. Publisher does not assume, and expressly disclaims, any obligation to obtain and include information other than that provided to it by the manufacturer.

The reader is expressly warned to consider and adopt all safety precautions that might be indicated by the activities herein and to avoid all potential hazards. By following the instructions contained herein, the reader willingly assumes all risks in connection with such instructions.

The publisher makes no representation or warranties of any kind, including but not limited to, the warranties of fitness for particular purpose or merchantability, nor are any such representations implied with respect to the material set forth herein, and the publisher takes no responsibility with respect to such material. The publisher shall not be liable for any special, consequential, or exemplary damages resulting, in whole or part, from the readers' use of, or reliance upon, this material.

ACKNOWLEDGEMENTS

The author and Delmar Learning wish to acknowledge and thank the following reviewers for their suggestions and comments:

Thomas Gustafson
Lake Superior College
Duluth, MN

Allen Sanderlin
York Technical College
Rock Hill, SC

Robert Jones
Independent Electrical Contractors
Houston, TX

CONTENTS

CHAPTER 4 Data Elements

CHAPTER 5 LAN Technologies

CHAPTER 6 Classic Ethernet

CHAPTER 7 Token Passing Rings

CHAPTER 8 Contemporary LAN Standards

CHAPTER 9 Internetworking Devices: Repeaters and Hubs

CHAPTER 10 Internetworking Devices: Bridges and Layer 2 Switches.

PREFACE

This book is the result of over two decades of experience in network computing. The information gathered and organized here was derived from my direct experience working for systems integrators and wholesale distributors, attending innumerable seminars, and reams of research conducted in the development of publications and course materials to facilitate teaching professional seminars since 1991.

The book is logically structured beginning with the basics of local area network technologies, a bit of LAN industry history, and an overview of networking standards, followed by a an explanation of the bits, bytes, frames and packets used by all networks.

As a basis for comparison all networking components will be defined according to their relationship to the ISO OSI Seven Layer Reference Model. This will help to provide an understanding of the function of each component and its relationship to other components.

Once the foundation in LAN technologies and components has been established, LAN standards are discussed in chronological order, beginning with Classic Ethernet. Token passing ring networks are next, including IEEE 802.5, IBM token ring, and FDDI/CDDI. The book then addresses more recent innovations in LAN technology: Fast Ethernet, Gigabit Ethernet and 10 Gigabit Ethernet. Extending the LAN infrastructure and internetworking multiple networks is discussed in subsequent chapters.

And since the most popular protocol in use on the planet is TCP/IP, the last chapter provides an in-depth tutorial on IP addressing, subnetting and supernetting. Because acronyms and confusing jargon are an inherent part of this industry, we will begin to tackle the arcane language of network computing right from the start.

CHAPTER OVERVIEW

Chapter 1 provides a LAN overview. The opening chapter builds a foundation of basic local area networking concepts by defining what LANs are, identifying their multiple uses, and explaining how they work.

Chapter 2 is a brief history of LANs. This chapter traces the evolution of LANs to help build an understanding of why we are where we are.

Chapter 3 explains the quagmire of standards bodies and organizations.

Chapter 4 discusses networking data elements: the bits and bytes of data transmissions. From binary to bytes, to frames, packets, segments and cells, each step in the data communications process manipulates different arrangements of these chunks.

Chapter 5 describes the network access protocols used by several of the most popular LAN technologies, their origins, and the standards that define each.

Chapter 6 is dedicated to Classic Ethernet; the original 10-megabit per second LAN technology initially defined by DIX ESPEC2 Ethernet and later standardized as IEEE 802.3.

Chapter 7 is dedicated to token passing ring LAN technologies including IEEE 802.5, IBM token ring, and FDDI/CDDI.

Chapter 8 covers contemporary LAN standards such as Fast Ethernet and Gigabit Ethernet.

Chapter 9 explains physical layer networking devices: repeaters and hubs.

Chapter 10 explains data link layer networking devices: bridges and layer 2 switches.

Chapter 11 explains network layer internetworking devices: routers and layer 3 switches.

Chapter 12 covers copper and optical fiber cable, cabling standards, and structured cable plants. The advantages and disadvantages of each type of media will be discussed, as well as the standards that have been developed to define them.

Chapter 13 provides a tutorial of IP addressing and subnetting. This final chapter covers Classful as well as Classless Internet Protocol addressing, how to create and identify Subnets and Supernets, and how to configure host systems for static or dynamic IP addressing.

LAN Overview

OBJECTIVES

After completing this chapter, you should be able to:

- Define what a local area network is.
- Identify the components of a LAN communications system.
- Describe the various purposes for using a LAN.
- Describe how LANs work.
- Identify the most common LAN technologies.
- Define the various LAN topologies.
- Identify the benefits of the star-wired topology.
- Better understand and use industry terminology.

INTRODUCTION

In the opening chapter we will start to build a foundation of basic LAN concepts by defining what local area networks are and identifying their multiple purposes. I will then provide a general understanding of how LANs work, and begin to explore the language we use to describe them.

WHAT IS A LAN?

One of the simplest definitions for local area network (LAN) is: a system to provide communications between multiple computers. Or, for greater detail and accuracy: a LAN is a communications system consisting of two or more computers in relatively close proximity, connected by a common medium, and using common protocols (see Figure 1.1). The next few paragraphs will elaborate on just what all this means.

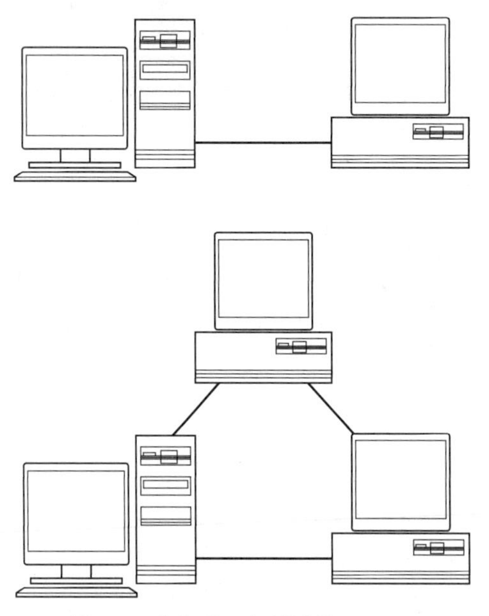

Figure 1.1 *Simple LAN Illustrations*
The simplest networks support just two computers connected back-to-back; adding more computers adds complexity.

The *communications system* consists of a network interface card (NIC) or network adapter in each computer that supports the same LAN technology, such as Ethernet or token ring. The *common medium* may be some form of cabling, such as coaxial, unshielded twisted pair (UTP), or optical fiber. The *common protocols* include the network access method used by the chosen LAN technology, such as Ethernet's carrier sense multiple access with collision detection (CSMA/CD) network access protocol, as well as the chosen network communication protocol, such as the Internet Protocol (IP) or Novell's Internetwork Packet Exchange (IPX).

Most LANs require a NIC or LAN adapter. Some computers provide a built-in LAN adapter whereas most require installing a NIC. A multitude of different LAN technologies has evolved over the years, including Ethernet, Arcnet, PC Net, token ring, LocalTalk, and fiber distributed data interface (FDDI). Each LAN technology requires all network-attached devices to have a network adapter or NIC that supports the appropriate media for that LAN technology. In other words, all NICs usually have to support the same type of media (coaxial, twisted pair, optical fiber, or wireless). Mixing different types of media is also possible with the appropriate networking hardware. It is also significant to note that some LANs are internationally accepted standards; along the way a few became de facto standards, and many have since become obsolete while others remain the proprietary technology of a single manufacturer.

As an aside, certain so-called zero slot LANs use the serial (COM) port or parallel (Printer) port commonly found on most personal computers. Since zero slot LANs are usually intended to support no more than two computers via a point-to-point connection, they are not considered a "shared media" network and are therefore excluded from further discussion.

LANs allow multiple computers to share a common medium and are therefore considered shared media networks. Shared media LANs can support communications between dozens or hundreds of computers, as well as provide point-to-point communications between just two computers. Because multiple computers have concurrent access to the shared medium, contention for access is certain to occur. To manage this contention, some mechanism must be provided to govern access to the shared medium. Several such mechanisms have been developed and they are referred to as "network access protocols," "network access methods," or "channel arbitration methods." Examples include carrier/sense multiple access with collision detection (CSMA/CD), used in Ethernet; and token passing, which is used in token ring and FDDI. Each station must be uniquely identified and some method of channel arbitration must be defined to govern access to the shared medium.

Over the years LANs evolved to support a variety of different media, ranging from copper wire to optical fiber, as well as radio wave and infrared. There is variety within each of these different types of media as well, and this topic will be covered in detail later in the book.

Protocols are everywhere in networking. The simplest definition for protocol in the context of networking is: a set of rules for communication. Some protocols complement other protocols and are grouped into "suites" (such as TCP/IP), other protocols are alternatives intended as outright replacements, and some protocols are required to support yet other protocols. For example, support of the network communications protocol TCP/IP on a shared Ethernet LAN requires the Ethernet network access protocol, CSMA/CD. Network communications requires many protocols working together to provide connectivity between two or more computers. The Transmission Control Protocol (TCP) requires the IP to support its functions, and IP relies on the Address Resolution Protocol (ARP) to resolve IP addresses to hardware addresses. There are many such interdependencies between network protocols.

PURPOSE OF A LAN

There are several possible purposes for local area networking. Some of the most obvious are to share expensive peripherals such as disk drives, archival (back-up) systems, printers, modems, and Internet access. Another typical purpose of LANs is to provide common access to shared data and program files. LANs can also be used to support multi-user programs, electronic mail, instant messaging, and powerful collaborative groupware applications (as well as multi-user games!). Without LANs these last functions are not possible.

HOW LANS WORK

Computer data and programs are stored and manipulated in binary form. One character of data typically consists of 8 bits, referred to as a byte. Data and programs are arranged in files consisting of some number of bytes. To move a data file or program file from one computer to another over a network, network communication protocols segment the files into smaller, more manageable blocks of bytes, usually called a *segment.* A communication protocol adds a header to the data segment, which typically provides application identification information, flow control, and finally error-checking information. The combination of data plus protocol header is referred to as a protocol data unit (PDU), and is passed down to the network protocol. The network protocol adds another header to the PDU, which typically provides network addressing, protocol identification information, and more error-checking information. The result is a network PDU, commonly called a *packet.*

To transmit packets of data and communication protocols over a network such as a LAN, a LAN access protocol is required. Each LAN access protocol encapsulates the packets inside a LAN frame. Like a packet, a frame consists of bits and bytes of information arranged in a specific order. The LAN frame and its contents are encoded electrically (or optically) for transmission over the common medium as a stream of bits. The bit streams are actually transmitted onto a common copper medium as voltage fluctuations. Each voltage fluctuation, or a group of voltage fluctuations referred to as "symbols," represents one or more bits depending on the data-encoding scheme being used (such as Manchester, 4B5B, or 8B10B encoding).

As a bit stream travels along the medium, it weakens, or attenuates, and both the wave shape and its timing become distorted. The physics of each type of media impose certain constraints on the length of that medium and the potential *bandwidth* a given media can support. To travel over greater distances, the voltage fluctuations must be reinforced and the distortions removed before the signal deteriorates beyond recognition. Networking devices that provide this functionality are called repeaters. Repeaters *retime* and *regenerate* weakened and distorted bit streams, restoring them to their original strength, timing, and shape. As the name implies, retiming corrects the timing distortion, whereas regeneration refers to reinforcing (or amplifying) the signal's strength.

The term *amplify* is typically used to refer to an analog function. Some older networks were in fact based on analog technologies, and as such used amplifiers to strengthen signals weakened from traveling over long distances. Amplifiers boost the entire spectrum of a transmission—the data as well as the noise. After a signal passes through enough amplifiers, the noise eventually becomes as "loud" as the data signal. In other words, the signal-to-noise ratio is diminished below system tolerances, and at some point the data can no longer be distinguished from the noise. By contrast, LAN repeaters are digital devices that read a deteriorated bit stream (a digital signal), and then correct the timing and retransmit that digital signal at 100% of its original strength. Unlike amplifiers, digital repeaters do not boost or propagate noise.

Data encoding refers to the various methods of arranging bits to represent the alphanumeric character set (A, B, C - 1, 2, 3), punctuation, and some special characters. The alphanumeric characters are encoded in one of several binary encoding schemes, such as ASCII (7 bit), EBCDIC (8 bit), or UNICODE (16 bit). The original ASCII code was based on 7 bits but most systems today will support an extended 8-bit ASCII character set. Each of these encoding schemes is simply an arbitrary arrangement of bits in groups of 8 (or 16 in the case of UNICODE) that represents the alphanumeric character set of a given language and its punctuation, as well as some special control codes such as carriage returns and line feeds.

ASCII	American Standard Code for Information Interchange (ANSI)
EBCDIC	Extended Binary Coded Decimal Interchange Code (IBM)
UNICODE	Required to support languages with very large character sets

CLIENTS, SERVERS, AND HOSTS

Like so many other terms in this industry, clear distinctions between *workstations, servers, clients, hosts, nodes* and *stations* tend to be elusive. Depending on the network technology being discussed, servers are often referred to as hosts, and nodes as stations. For the sake of consistency and to avoid confusion, the definitions used by widely recognized standards bodies are adhered to throughout this text (see Figure 1.2).

The term *station* is generally used to refer to any computer attached to a network regardless of its make or size, such as a personal computer or a mainframe. The term *node* generally refers to circuit switching points within a network. Although most network switches employ many of the same components as computers (processor, memory, etc), switches are not used for computing per se.

The term *server* refers to a network-attached computer that provides access to one or more shared resources. There are many kinds of servers, including file servers, printer servers, application servers, terminal servers, communication servers, and fax servers. As the names imply, file servers provide shared access to a common file repository, usually via one or more large disc drives; and printer servers provide access to one or more shared printers. Application servers are a bit more complicated, but they typically consist of a combination of computer hardware and an operating system optimized to support certain multi-user application programs, such as a database.

Whereas terminal servers are specialized network devices that connect a number of so-called *dumb terminals* to a LAN, communications servers are usually network-attached

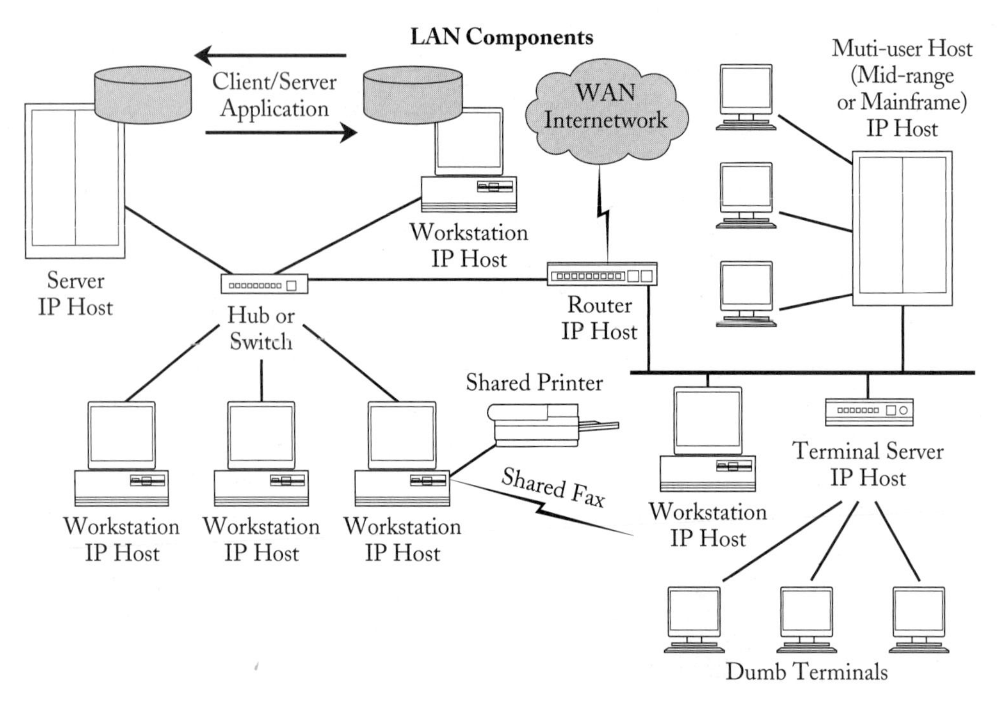

Figure 1.2 *Workstations, servers, clients, hosts, and stations*

computers that connect a LAN to one or more other network services, such as a shared pool of dial-up modems, access to private data networks, or access to the public Internet.

Another type of communication server is the fax server, which allows all of the computers on a network to share a common facsimile resource. What makes any network-attached computer a *server* depends on the software running on that computer, and how that software is configured. The term *workstation* refers to a network-attached computer on which users run different application programs to do their work. By installing special software, or configuring certain features of the operating system, any workstation can support a variety of server functions. Common server applications found on workstations include file and printer sharing, fax sharing, and Internet access sharing.

The term *host* has two common meanings. In Internet Protocol (IP) parlance, *host* may be used to refer to any computer, workstation, server, client, station or node with an IP address. In a non IP-specific context, a *host* may be any network-attached multi-user computer, such as a mainframe, mini- or mid-range computer.

The term *client* generally refers to a workstation that uses the shared resources of server computers. *Client/Server* refers to a software architecture that divides program duties between the server computer and each client workstation. The specific details of this division of duties vary from vendor to vendor, thereby generating even more confusion for everyone.

DEFINING THE TOPOLOGY

The uses of the term *topology* can be confusing and contradictory. It is therefore necessary to first identify the contexts in which the term is to be used. Only then can we define the meaning of the word topology (see Figure 1.3).

Physical describes the actual layout or path of the media or cabling installed in a building or campus, or the appearance of the cable runs on a blueprint or floor plan.

Electrical describes the behavioral characteristics of the signal as it traverses the media, such as that found in a broadcast bus environment or a circulating ring environment.

Logical more accurately describes the functional characteristics of the network access method or network access protocol used, e.g., CSMA/CD or token passing.

The traditional topology for Ethernet, for example, is both a physical and electrical bus. However, 10Base-T Ethernet is a star-wired bus—physically the cabling is installed in a star but electrically it still behaves as a bus. For comparative purposes, refer to star-wired Ethernet as a hybrid topology because it combines the physical characteristics of a star and the electrical characteristics of a bus.

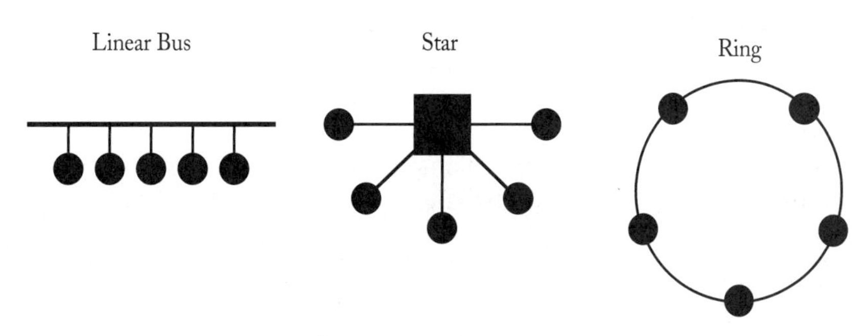

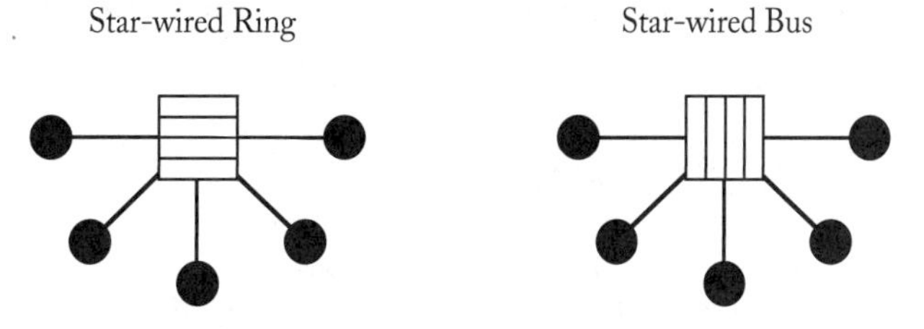

Figure 1.3 *Network Topologies*
Over the years the simpler *traditional* topologies have been largely displaced by new *hybrid* topologies.

Some may refer to their LAN as having an Ethernet topology, but because Ethernet supports several different physical topologies this can only infer the CSMA/CD network access protocol, which functions electrically as a bus. Although the network access protocol may be known, you still cannot be certain as to the network's physical topology. Recent advancements in networking technology have developed efficient, flexible combinations of what were once simple, rigidly defined networking architectures. The simpler traditional topologies have been largely displaced by new hybrid topologies. For example, 100Base-T Fast Ethernet supports only UTP or optical fiber media configured as a star-wired bus.

TRADITIONAL LINEAR BUS

The coaxial linear bus is the oldest and at one time was the dominant Ethernet topology in use. This is the topology defined by the original specifications of both DIX (Digital Intel Xerox) and IEEE (Institute of Electrical and Electronic Engineers). The first IEEE 802.3 Ethernet specification included this topology over thick coaxial media (or coax—pronounced *"kō'-ăks"*), which is today identified as 10Base 5 (see Figure 1.4).

Later the 802.3 specifications were appended to support thin coax, also called *thinnet* or *cheapernet.* Identified by the IEEE as 10Base 2, this standard states that RG-58a/u or RG-58c/u thin coaxial cable satisfies the electrical requirements of the specification. This version of Ethernet is virtually identical to 10Base 5, in that both operate at 10 Mbps, use CSMA/CD, support a linear bus topology, and require a 50-Ohm terminating resistor at both ends of each segment. The major differences between them are the length of the linear bus segment and the number of transceivers that can be attached to each coaxial bus segment. 10Base 2 supports shorter segments and fewer transceiver connections than 10Base 5, but as the nicknames imply, the cabling used for 10Base 2 is thinner, making it easier to install and much less expensive than the thick coax required by 10Base 5.

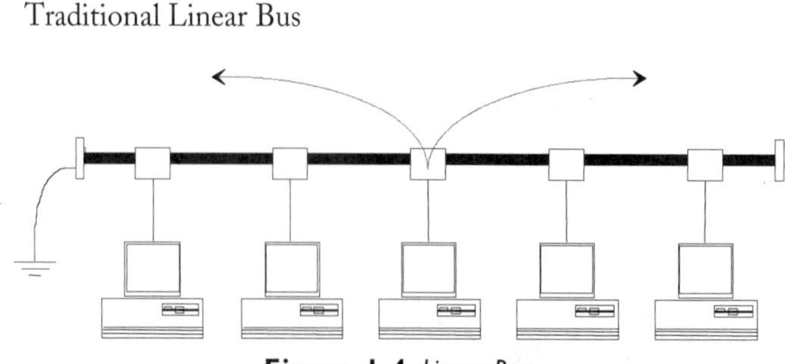

Figure 1.4 *Linear Bus*
Physical bus, electrical bus, logical bus (CSMA/CD);
IEEE 802.3 10Base 5 using thick coaxial media (Ethernet).

All devices (computers, repeaters) connect along the length of the bus. Both ends of the bus must be properly terminated. Coaxial Ethernet requires a 50-Ohm terminating resistor at each end. The media may be coax (as in certain Ethernet implementations) or UTP (as in Apple Computer Corporation's LocalTalk). One end of each thick coax segment must be earth grounded. (The specifications actually call for grounding at *one point* of each segment, but it is almost always preferable to ground one end at the terminator.) Many thick coax (N-Series) terminators provide a grounding clip or lug. Thin coax should also be grounded at one point, although the majority of all thin coax installations are not earth grounded at all. LocalTalk and most other LANs do not require special grounding.

The local area network medium provides a shared transmission channel. Because multiple computers share the transmission channel, some method of controlling access is required. Channel arbitration in physical bus networks may be governed by a contention protocol such as CSMA/CD (Ethernet), CSMA/CA (LocalTalk), or a deterministic token-passing bus protocol such as that used by the now antiquated Arcnet. The most common of these, CSMA/CD, permits any station to access the shared medium anytime the medium is not currently in use by another station. More detail on CSMA/CD will be provided in a subsequent chapter.

TRADITIONAL RING

This topology is usually the least understood and most poorly explained. Virtually all ring networks utilize some form of token-passing network access protocol. Early

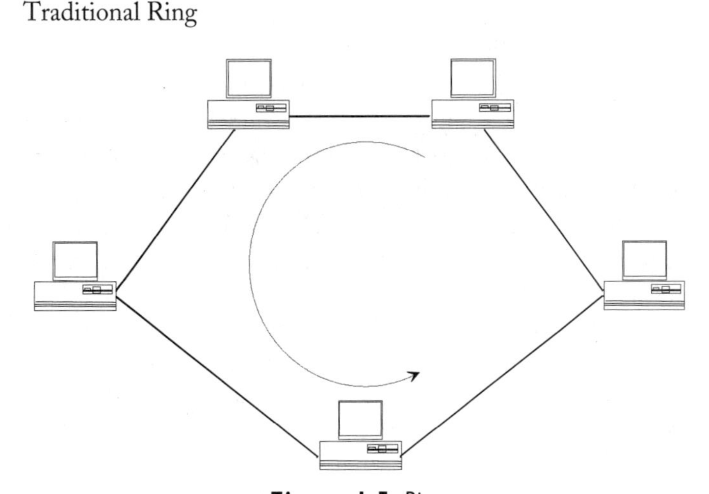

Figure 1.5 *Ring*
Physical ring, electrical ring, logical ring (token passing)
abstract of IEEE 802.5 and ANSI X3T9 FDDI (token ring)

ring technology actually did daisy chain from device to device to form a complete circle. Most of today's commercially available ring networks do not physically resemble this picture. Two of the earliest such ring networks developed were the Zurich Ring and the Cambridge Ring. In this configuration, a fault in any station, or anywhere in the media, would be catastrophic, causing total network failure. This illustration is NOT representative of the IEEE 802.5 specifications or IBM Token Ring networks (see Figure 1.5).

Channel arbitration in physical ring networks is always controlled by a token-passing protocol such as that used by IEEE 802.5 token ring and ANSI X3T12 FDDI (formerly known as ANSI X3T9). Tokens are specialized frames that constantly circulate from station to station around the ring to arbitrate, or govern, access to the shared medium. All stations must wait for a *free token* to gain access to the medium. Once access to the medium has been acquired, the station may transmit for an amount of time determined by token ring timing mechanisms.

TRADITIONAL STAR

Few early LANs were actually based on a star topology. The example of the terminal/host network provides only slow serial connections to dumb terminals using interfaces such as RS-232 or RS-423. The top speed of these links was typically 9600 bits per second, or 19,200 bps in some cases. These speeds are not considered adequate for the distributed processing nature of LANs, nor today's rapidly growing bandwidth demands (see Figure 1.6).

Other terminal/host systems that employed a traditional star network such as Unix hosts and those from Digital Equipment Corporation (DEC—acquired by Compaq, which is now owned by HP.) directly support Ethernet and other LAN technologies. As the organizations using these systems grew, demands on these host computer systems increased, and more and more terminals were wired into the host. The tangle of wiring became unmanageable, and a solution emerged that provided computer terminals access to their host indirectly via terminal servers that could be distributed along the length of a LAN segment. An interface was installed in the host computer to provide its attachment to the LAN and another in the terminal servers. The LAN connection supported greater distances, numerous terminal servers, and therefore greater numbers of terminals. Special protocols were developed to support communications between terminals and hosts. In addition to developing the host computers themselves, most computer vendors developed their own protocols, terminals, operating systems, and other software applications. Once LAN interface cards were developed for personal computers, the *decade of the LAN* was launched.

STAR-WIRED BUS

This is the definition of a star-wired network in today's terms. Like the traditional bus, the star-wired bus is a *broadcast* environment where all devices *hear* all transmissions

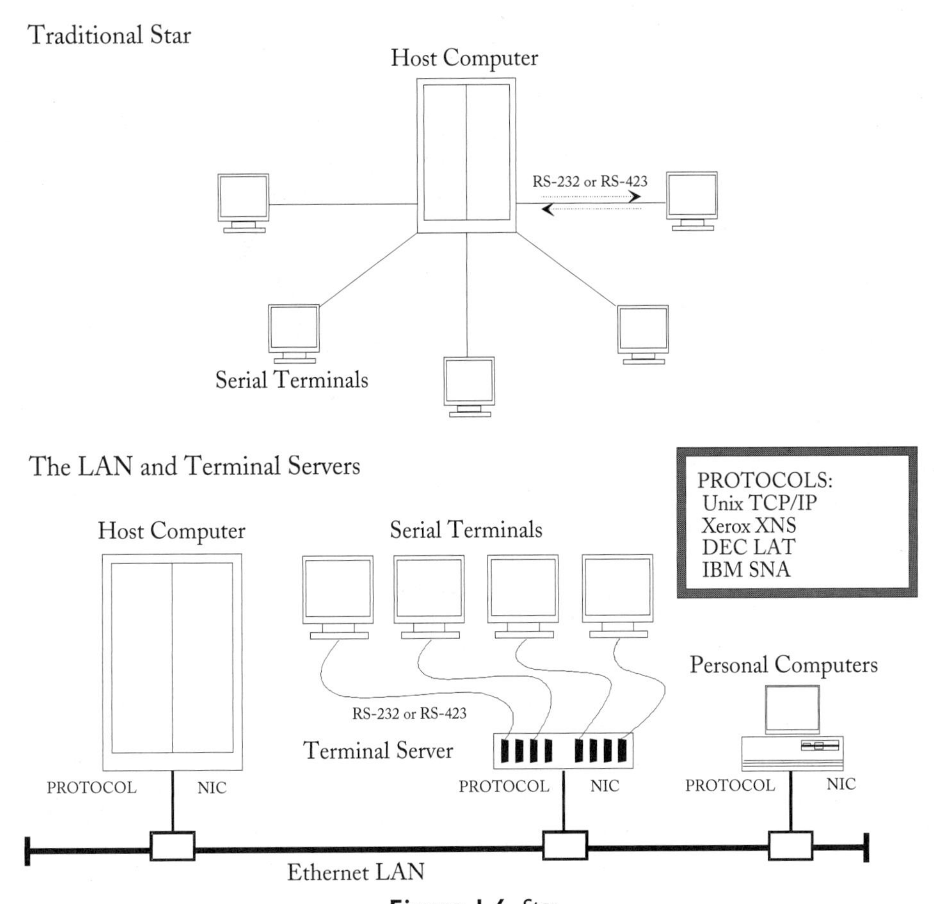

Figure 1.6 *Star*
Physical star, electrical star, logical star (various protocols);
illustrates early terminal/host-based networks and the evolution to support LANs.

from any device simultaneously. In contrast to the traditional star, the *central unit* or hub is NOT a host computer or file server, although many manufacturers have included some level of intelligence in their hubs to provide advanced management, fault isolation, and notification capabilities. The 10Base-T *central unit*, hub or concentrator, is a *multi-port repeater*. All ports on the hub are considered to be one *repeater-hop* from all other ports on the same hub (see Figure 1.7).

Each station is connected to the hub by its own UTP cable run, called a *link segment*. Link segments require two pairs of conductors, one pair to transmit and one pair to receive. The 10Base-T specification was designed to support ordinary DIW-24 UTP telephone cabling to a distance of at least 100 meters. In the early days of 10Base-T development, several vendors offered support for distances up to 150 meters using what was then referred to as *high-performance* UTP cabling.

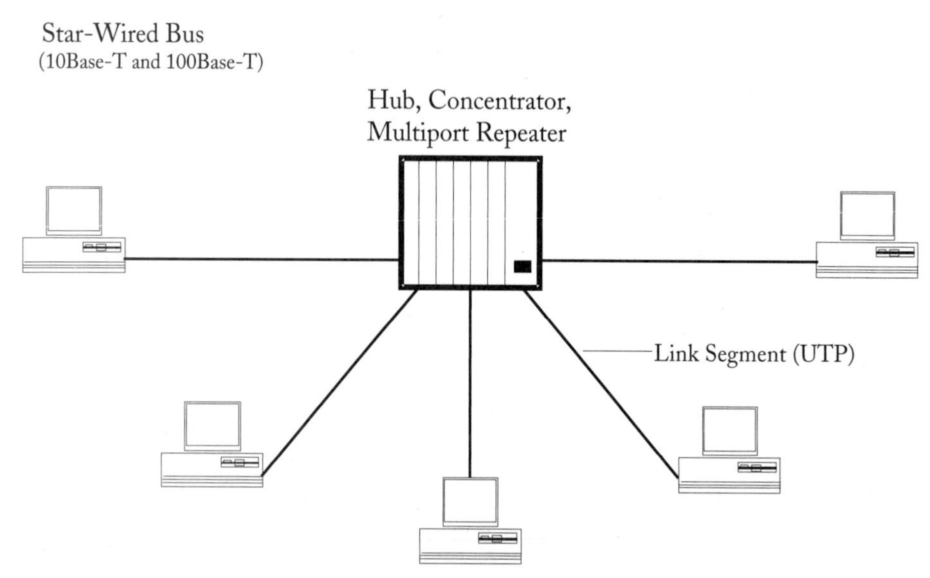

Figure 1.7 *Star-Wired Bus*
Physical star, electrical star, logical bus (CSMA/CD);
IEEE 802.3 10/100Base-T using UTP media (Ethernet).

The physical topology extent is determined by the Ethernet repeater rules and the media's round-trip propagation delay. The repeater rules differ between 10Base-T Ethernet and 100Base-T Fast Ethernet, and using optical fiber can extend their physical topologies. The maximum length of a UTP link segment for both 10Base-T and 100Base-T is 100 meters. The specifics of each of these standards will be covered in their respective chapters.

Benefits of the star-wired bus topology include:

Modular design, which provides a scalable network architecture that can be tailored to satisfy virtually all data cable plant requirements.

Cost savings achieved by simplifying moves, adds, and changes, and reducing network downtime due to cable plant faults.

Automatic connection partitioning on a port-by-port basis isolates any faulty areas while continuing to provide service to unaffected portions of the network.

The IEEE 802.3i 10Base-T specification requires auto-partitioning and auto-reconnection. The latter means that once a fault is corrected, communications through the port will be resumed dynamically—no user intervention is required.

STAR-WIRED RING

This is also a definition of a star-wired network. In contrast to the early versions of the traditional ring, a central hub or Multi-Station Access Unit (MSAU) is required. This centrally located MSAU (or MAU) virtually eliminates potential catastrophic failures by providing a means of circumventing most points of media failure as faults occur. This is accomplished by using a by-pass relay at each *lobe port* (station connection) on the MSAU. Additionally, a few manufacturers have included some level of intelligence in their MSAUs to provide advanced management capabilities including activity monitoring, fault isolation and notification, beacon recovery, fault tolerance, as well as support for multiple different network technologies in a single chassis (see Figure 1.8).

Each station is connected to the MSAU by its own UTP or shielded twisted pair (STP) cable run, in this case called a *lobe*. As with link segments, lobes require two pairs of conductors, one pair for transmit and one pair for receive. The original IEEE 802.5 specification supported only 4 Mbps operation using only the more expensive and bulky (150 Ohm) STP. Distances were determined based on a complex formula consisting of numerous variables. Support was later added for 16 Mbps operation and *high-performance* (100 Ohm) UTP cabling, which later became known as TIA/EIA Category 4. More detail on cabling categories will be provided in a subsequent chapter.

Using STP cabling, the physical topology extent is determined by the type of media, length and number of lobes, type and number of MSAUs, whether the network is operating at a data rate of 4 Mbps or 16 Mbps, maximum frame size (which varies based on the data rate), inter-MSAU cables, and the limitations of the *Token Rotation Timer*. Although STP is technically a superior media, it is more expensive not only to

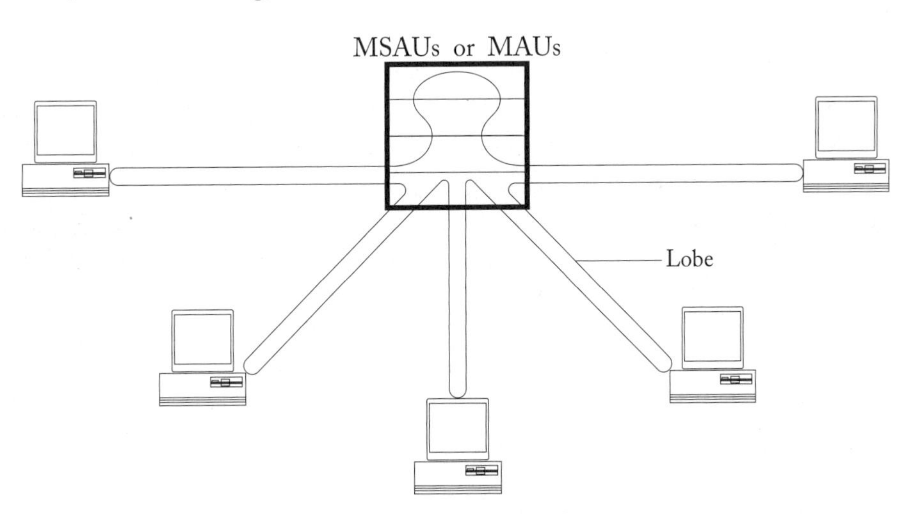

Figure 1.8 *Star-Wired Ring*
Physical star, electrical ring, logical ring (token passing);
IEEE 802.5 using UTP or STP media and a stack of four MSAUs (token ring).

purchase but also to ship, install, and maintain. It is much heavier and bulkier than UTP, and if the connectors are not properly installed, the shielding can cause more problems than it was intended to prevent.

Using UTP cabling, the physical topology extent is limited by the electrical characteristics of the media, which is usually 100 meters (328 feet). Again, using optical fiber can be used to extend the physical topology. The specifics of each of these standards will be covered in their respective chapters.

Benefits of the Star Wired Ring topology include:

Modular design, which provides a scalable network architecture that can be tailored to satisfy virtually all data cable plant requirements.

Cost savings achieved by simplifying moves, adds, and changes, and reducing network downtime due to cable plant faults.

Automatic connection partitioning on a port-by-port basis is designed to isolate faulty areas while continuing to provide service to unaffected portions of the network. This feature is provided via the bypass relay of each lobe port.

BUZZWORDS AND TLAS

Networking terminology has become increasingly confusing in recent years. Different vendors, authors, and users use technical terms interchangeably. Definitions for some of these terms have been standardized by various standards organizations such as the IEEE, ANSI, ISO, ITU-T, EIA, TIA, and the IETF. Many vendors have developed products that incorporate the functionality of two or more discreet internetworking devices into a single product. In an attempt to name these products appropriately, some vendors resort to coining their own terms. Other vendors then borrow these terms and apply them to their own products that have similar yet quite different capabilities. As a result, many different products share similar names, which causes confusion for users trying to make networking product selection decisions.

One such problematic term is *brouter*. Obviously, this refers to a device that is both a bridge and router. It becomes less obvious when one considers that there are several different methods of bridging, and at least eight different routable protocols, several of which support more than one routing update protocol. This little bit of information prompts an array of questions that must be answered before one can determine if a given brouter actually satisfies the networking requirement. The only thing obvious about the term brouter is that it is a marketing term still found in some vendors' *brochureware*.

Another technical term with many meanings is *switch*. Depending on one's background or the context in which the term is used, switch can mean many things. For example, in a telephony context, it may refer to a private Branch Exchange (PBX) that switches physical circuits. In the context of Asynchronous Transfer Mode (ATM), it refers to an ATM cell switch that relies on virtual circuits. And in the context of LANs, it may refer to any of several devices, including a multi-port bridge that switches frames, or a multi-port router that switches packets.

Technical jargon, buzzwords, and three-letter acronyms (TLA) are pervasive in the computer and networking industries. The alphabet soup can be overwhelming even when one knows most of their meanings! Things have gotten so bad that the same TLAs appear in different facets of the computer and networking industries, but with different meanings. Using these terms properly serves to convey competence and failure to do so conveys the inverse. Figure 1.9 groups many common networking terms and acronyms by their relationships rather than alphabetically.

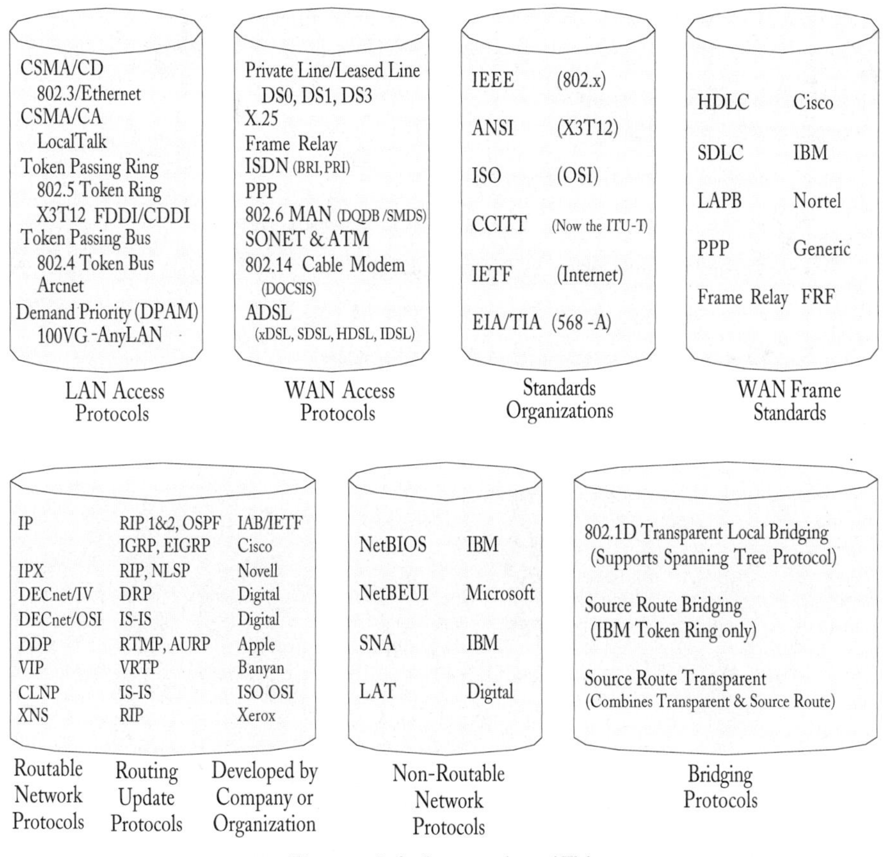

Figure 1.9 *Buzzwords and TLAs*
The networking industries are replete with buzzwords, arcane jargon, and myriad three-letter acronyms (TLAs).

CHAPTER SUMMARY

Many advantages are gained by networking computers, including the ability to share expensive resources, access to multi-user applications, and the ability to instantly exchange messages and files with other network users. LANs require a NIC in each computer, a common medium, a communications protocol; a hub is usually required with three or more computers. Many different and varied approaches to networking have emerged over the years, but only a few have survived. Most notable of these is Ethernet, which has evolved to support the demand for greater speeds and flexibility. And finally, the industry is littered with acronyms and confusing jargon essential to communicating complex concepts. It is essential to use technical terms properly.

REVIEW QUESTIONS

1. What is a local area network?
2. List the components of a LAN communications system.
3. List the various purposes for using a LAN.
4. LANs transmit information is what form?
5. What is the most common LAN technology?
6. Which LAN employs a star-wired bus topology?
7. Identify the benefits of the star-wired topology.
8. Define the following acronyms:

NIC	TLA
IP	IPX
ARP	IEEE
ASCII	EBCDIC
UTP	DIX

9. What is the difference between an electrical and a physical topology?
10. Which LAN uses CSMA/CD?

A Brief History of LANs

OBJECTIVES

After completing this chapter, you should be able to:

- Describe the origins and evolution of Ethernet.
- Identify several of the key IEEE standards for Ethernet.
- Compare and contrast various LAN technologies.
- Identify where Ethernet was invented.
- Identify where token ring was invented.
- Explain why token ring succeeded.
- Explain why token ring failed.
- Explain why Ethernet prevailed.
- Define baseband transmissions.
- Define broadband transmissions.

INTRODUCTION

It is easier to understand where you are going if you already know where you've been. To gain an understanding of where the LAN industry is headed, you first need some knowledge of its history and evolution. This chapter traces the evolution of LANs to help build an understanding of why we are where we are. Throughout the chapter I will attempt to provide a unique perspective on the evolution of the technologies and the organizations responsible.

LAN EVOLUTION

Since the introduction of the first Ethernet LAN, several different types of LANs have been developed. Each provided different benefits in price/performance, media, and/or topology. Price/performance refers to value, as in getting the most "bang for the buck," or in the case of LANs, bits per second per dollar. Different media provides different benefits, and usually defines the topologies supported. Ethernet was developed at the Xerox Palo Alto Research Center (Xerox PARC) and was based on the "Aloha" protocol. Aloha formed the foundation of the carrier sense multiple access (CSMA) network access protocol. The name *Ethernet* is capitalized because it is a registered service mark of the Xerox Corporation. One definition for the network's rather odd "ether" reference is as follows: *Ether (Physics.) An all-pervading, infinitely elastic, massless medium formerly postulated as the medium of propagation of electromagnetic waves.*

Ethernet was first introduced as a 1 to 20 megabit per second (Mbps) LAN. The first commercial Ethernet hardware produced was based on a single type of thick coaxial cabling and operated at 10 Mbps. Since its inception Ethernet has evolved to support different types of media and ever-increasing speeds, and has become an international standard and the dominant LAN on the planet. Once Ethernet was standardized by the IEEE as 802.3, the first minor change was to add support for a less expensive and easier to use thin coaxial cabling. This was soon followed by the addition of optical fiber, which supports much greater distances, superior security, and noise immunity.

Next, in what was perhaps Ethernet's most significant evolutionary step, support was added for a telephony-based structured cabling plant using ordinary unshielded twisted pair (UTP) telephone wire. In this configuration, each LAN-attached computer gets its own dedicated cable run back to a central hub or concentrator. The central hub or concentrator is a multi-port repeater; like any other Ethernet repeater, hubs provide bit retiming and regeneration. Any bit stream received on any hub port is propagated out to all other hub ports. This media adaptation secured Ethernet's role as the most popular LAN technology for years to come (see Figure 2.1).

The next evolutionary step for LANs was the development of the LAN switch. LAN switches are, in essence, very fast, inexpensive multi-port bridges with very low latency. Unlike multi-port repeaters that forward bits, switches and bridges forward whole frames. Furthermore, most switches and all bridges forward frames based on the hardware (or station) addresses found in the headers of those frames. The benefit of this selective forwarding is to keep traffic off LAN segments where it does not belong. Whereas repeaters forward all bit streams onto all LAN segments interconnected by repeaters, switches and bridges forward frames only onto LAN segments where the destination station resides. These internetworking issues will be covered in much greater detail in the next chapter.

Increasing the speed from 10 Mbps to 100 Mbps (Fast Ethernet), and then to 1000 Mbps (Gigabit Ethernet) gave Ethernet a lock on the LAN market that it will likely hold indefinitely. A 10,000 Mbps (10 gigabit per second) Ethernet implementation is also now available and work is under way to develop 40 Gbps and 100 Gbps solutions based on Ethernet.

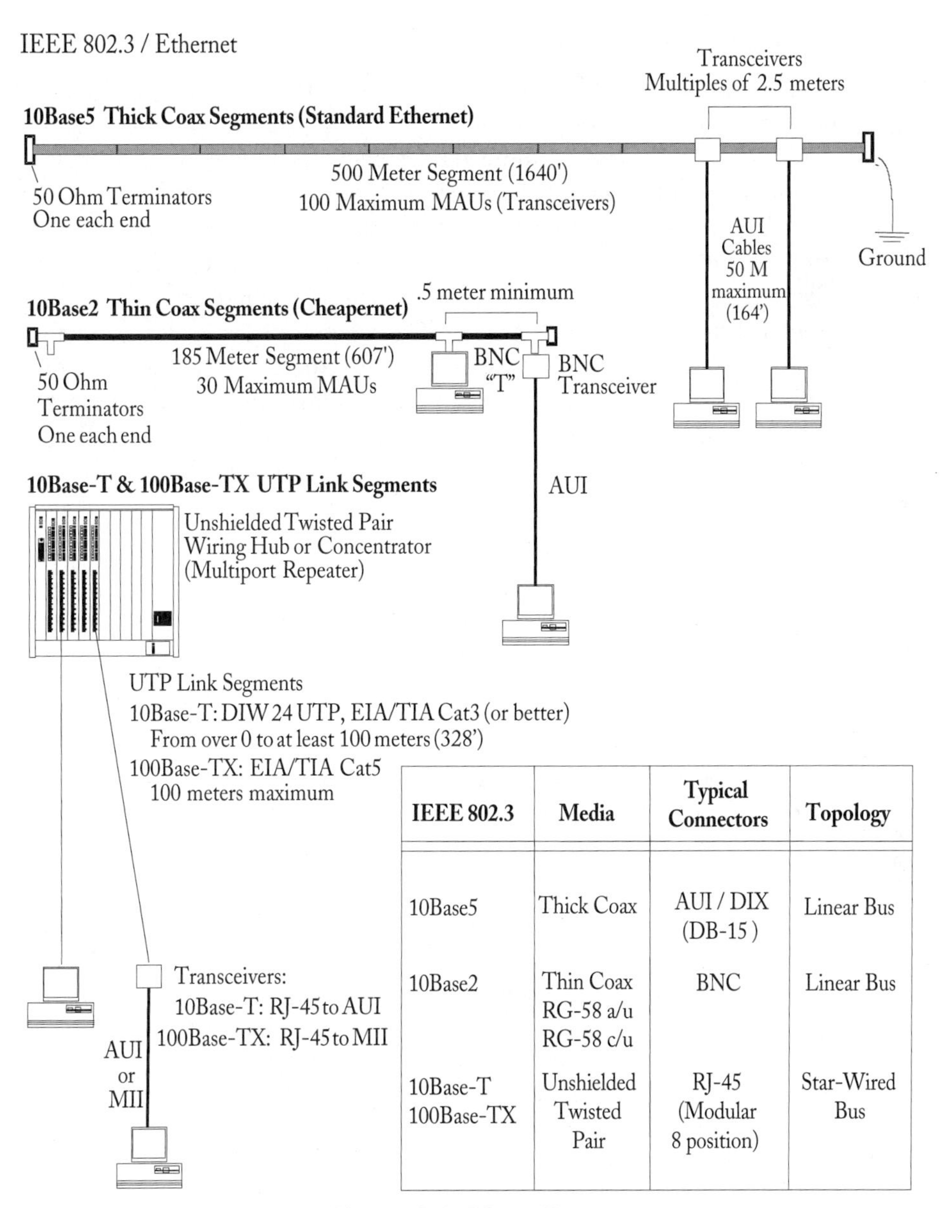

IEEE 802.3	Media	Typical Connectors	Topology
10Base5	Thick Coax	AUI / DIX (DB-15)	Linear Bus
10Base2	Thin Coax RG-58 a/u RG-58 c/u	BNC	Linear Bus
10Base-T 100Base-TX	Unshielded Twisted Pair	RJ-45 (Modular 8 position)	Star-Wired Bus

Figure 2.1 *Ethernet Versions*
Ethernet supports a wide variety of different types of media, each of which supports different topologies and imposes different distance limitations.

Table 2.1 summarizes the history of Ethernet.

During the early years of LAN development Ethernet had a lot of competition, most notably Arcnet. IBM Token Ring Network (TRN) did not appear on the scene until late 1985. During the early to mid-1980s Arcnet was an extremely popular and much more economical alternative to Ethernet. Key to its popularity was the fact that Arcnet sold at a fraction of the price of Ethernet (hundreds of dollars less). Pioneered by Pure Data, Arcnet achieved de facto standard status when companies such as Standard Microsystems Corporation (SMC) embraced the technology. Like Ethernet, Arcnet began with support for coax and was later adapted to use UTP. Arcnet supported both a bus topology and the more-popular hub-based distributed star topology, and employed a token-passing bus network access method. Unlike Ethernet, the speed of Arcnet remained at a relatively slow 2.5 Mbps until it finally dropped off the radar screen altogether. Although a 20 Mbps version did appear in Arcnet's final days, it never took off.

Before IBM Token Ring there was the IBM PC Net, an extremely unreliable 1 Mbps coaxial cable-based broadband LAN technology that was plagued by a plethora of problems few could resolve. Fortunately PC Net died before it did too much damage. Although Token Ring is vastly superior to PC Net, most consider it an overengineered and overpriced alternative LAN that is difficult to maintain and troubleshoot.

Standardized by the IEEE as 802.5, the 4 Mbps version of IBM Token Ring appeared at the end of 1985. It was an immediate success with IBM's customers, primarily because, aside from PC Net, it was the only LAN that IBM really endorsed at the time.

TABLE 2.1 THE HISTORY OF ETHERNET

Year	*Organization*	*Standard*	*Description*
Pre-1980	Digital, Intel, Xerox	ESPEC1 & 2	10 Mbps Ethernet (CSMA/CD) over coax
1980	IEEE 802.3	10Base5	10 Mbps CSMA/CD over thick coax
1982	IEEE 802.3	10Base2	10 Mbps CSMA/CD over thin coax
1987	IEEE 802.3i	10Base-T	10 Mbps CSMA/CD over DIW 24 UTP
1995	IEEE 802.3u	100Base-TX	100 Mbps over 2-pairs of Cat5 UTP
1995	IEEE 802.3u	100Base-T4	100 Mbps over 4-pairs of Cat3 UTP
1998	IEEE 802.3z	1000Base-SX	1000 Mbps over multi-mode fiber
1998	IEEE 802.3z	1000Base-LX	1000 Mbps over single-mode fiber
1999	IEEE 802.3ab	1000Base-T	1000 Mbps over 4-pairs of Cat5/Cat5E UTP
2002	IEEE 802.3ae	10GBase-X	10000 Mbps over multi-/single-mode fiber

At first the only media supported was the relatively expensive, bulky, and difficult to install IBM Type-1 shielded twisted pair (STP). STP can support greater distances and provide superior noise immunity over that of UTP; however, if the shielding is not grounded properly it can cause more problems than it is supposed to eliminate. As with Ethernet, the addition of optical fiber soon followed the copper implementations.

From 1985 until the mid-1990s token ring was very popular, gaining as much as 35% of the LAN market, even though it has always been a much more expensive alternative to Ethernet. And although token ring did achieve some commercial success under IBM's guidance, it was actually invented elsewhere. Credit for token ring's invention goes to a fellow named Olaf Soderblum, who developed a token-passing ring network to connect IBM mainframes in Sweden. IBM liked the technology and subsequently licensed it from Mr. Soderblum. IBM's development efforts took token ring from Sweden to Zurich and finally to the United States, where it was submitted to the IEEE for consideration as a standard.

Like Ethernet, token ring began with support for a clumsy cabling system and was later adapted to use UTP. Although both the IBM and IEEE specifications for token ring call for a star-wired topology, other vendors such as 3Com offered token ring products that supported an unpopular pseudo-bus topology.

Third-party token ring vendors were the first to offer support for the less expensive and easier to use UTP cabling. Although STP does support a structured cable plant, customers really wanted a telephony-based structured cabling plant that used ordinary UTP telephone wire. As with Ethernet, in this configuration each LAN-attached computer gets its own dedicated cable run back to a central token ring hub called a Multi-Station Access Unit (MSAU or MAU).

Although the physical topologies for Ethernet and token ring over UTP appear to be identical, the electrical topologies are quite different. When an Ethernet hub receives a transmission from one station, it repeats that bit stream out to all other attached stations, including the destination station. Unlike Ethernet, the token ring hub is *not* a multi-port repeater, and it does *not* provide bit retiming or regeneration. MSAUs are passive devices that rely on mechanical relays at each port to complete the ring circuit. The computer attached to each MSAU port energizes that port's by-pass relay. Each token ring NIC acts as a regenerating repeater but does not perform bit retiming.

Bit streams are passed from one station's NIC to its MSAU port, from the next MSAU port to the next NIC, from that NIC back to its MSAU port, from the next MSAU port to the next NIC, and so on until the bit stream circulates around the entire ring, passing through all active computers attached to the LAN. When a token ring MSAU receives a transmission from one station, it forwards that bit stream out to the next station in the circuit. That station forwards the bit stream back to the MSAU, which in turn forwards the bit stream out to the next station in the circuit. When the destination station finally receives the bit stream it reads the contents of the frame and then forwards the bit stream back to the MSAU, which in turn forwards the bit stream out to the next station in the circuit. This process continues until the transmission returns to the source station, where it is then purged from the ring and a new free token is released. In essence, every transmission must pass through every potential point of failure in the token ring LAN (see Figure 2.2).

The 16 Mbps version of token ring appeared in 1988. Both the 4 Mbps and 16 Mbps versions support STP cabling, and later, in response to customer demand, support for UTP cabling was added. Also in response to customer demand, third-party vendors and IBM released intelligent token ring hub systems. IBM's intelligent token ring hub products were called Control Access Units (CAUs) and Lobe Access Modules (LAMs). A single CAU can support up to four LAMs, and each LAM provides 20 token ring lobe connections (20 token ring stations) (see Figure 2.3).

Sometime around 1998, an alliance of vendors including IBM, Madge, Olicom, and about forty other companies teamed to develop a 100 Mbps version of token ring dubbed *High-Speed Token Ring*. This technology failed to capture any significant

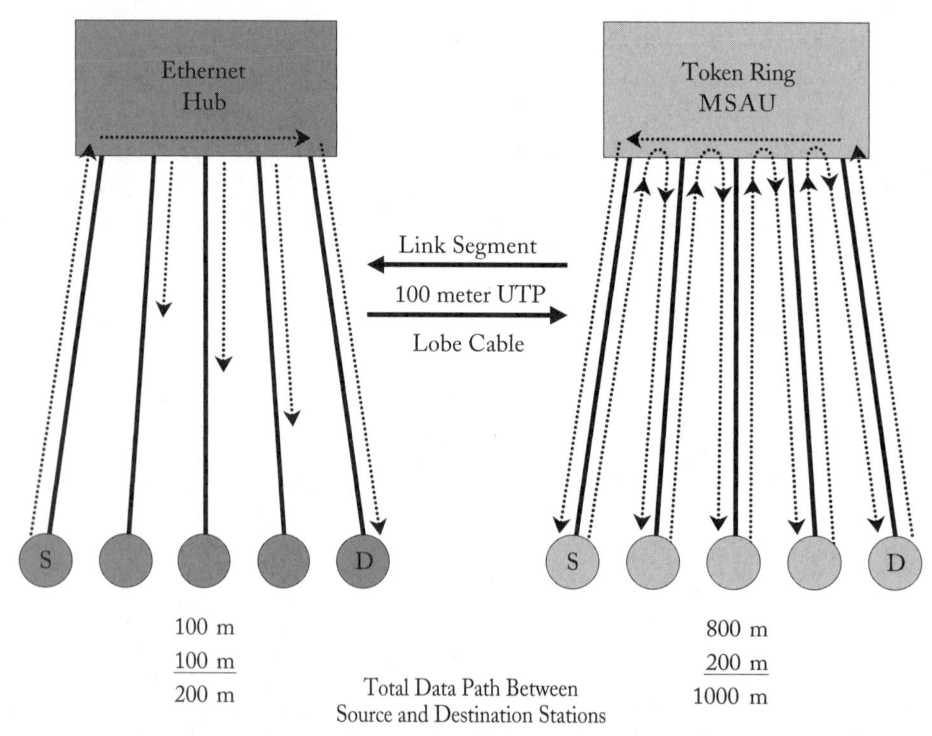

Figure 2.2 *Ethernet Hub vs. Token Ring Hub*
Ethernet transmissions pass from the source station to the hub, where the signal is simultaneously repeated out to all attached stations. In contrast, token ring transmissions circulate sequentially through each station.

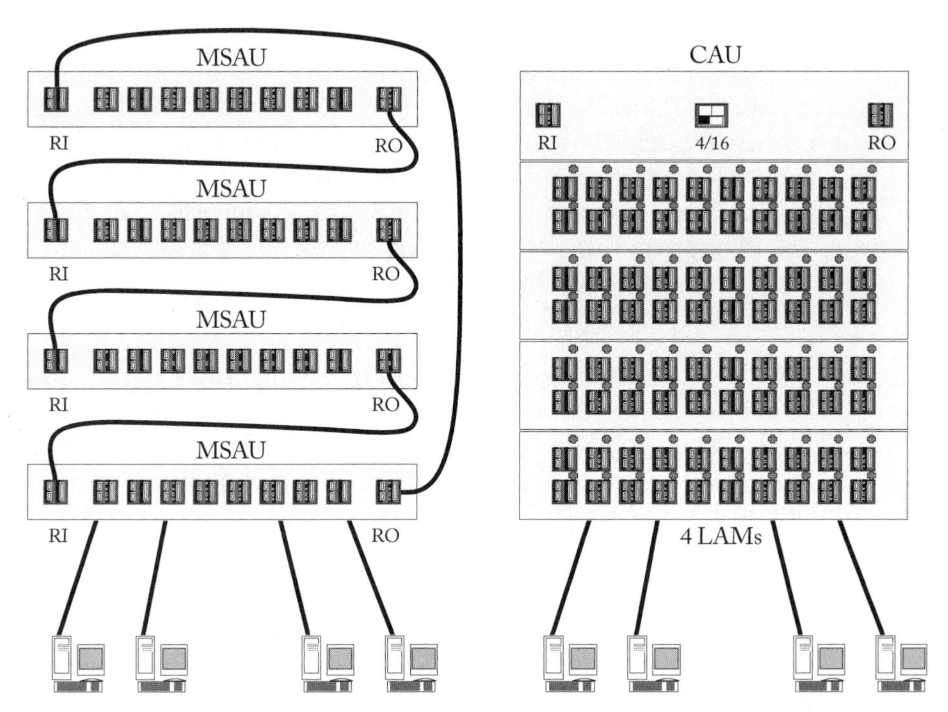

Figure 2.3 *Token Ring MAU, CAU, and LAM*
Multi-Station Access Units (MSAUs) provide 8 Lobe ports. One Control Access Unit (CAU) supports 4 Lobe Access Modules (LAMs), and each LAM provides 20 Lobe ports. Ring-In and Ring-Out ports interconnect MSAUs and CAUs.

market share. The most widely accepted upgrade path for 4/16 Mbps token ring is 100 Mbps Fast Ethernet and 1000 Mbps Gigabit Ethernet.

In 1987, over a decade before the appearance of 100 Mbps High Speed Token Ring, another 100 Mbps token ring LAN called Fiber Distributed Data Interface (FDDI) was released. As the name implies, FDDI supports optical fiber media. Specifically, FDDI is a 100 Mbps fault-tolerant counter-rotating fiber optic token-passing ring. Within a few years of its release, support was provided for shielded and unshielded twisted pair cabling, but most of these early copper implementations of FDDI were proprietary. Eventually everyone standardized on a set of specifications now called the Copper Distributed Data Interface (CDDI). While FDDI/CDDI did manage to gain as much as perhaps 12% market share at its peak, the technology has all but disappeared due to the continued high cost of the products. FDDI still exists in the backbone installations of some companies, university campuses, and Internet service providers.

Myriad other proprietary LAN technologies appeared and experienced some success, but were eventually overwhelmed by a few LAN standards. These include Corvus OmniNet, Gateway Communications G-Net, DCA (and later Fox) 10-Net, Western

Digital StarLAN (which became another IEEE standard—802.3 1Base5), Apple Computer LocalTalk, Proteon ProNET/10 and ProNET/80, to name but a few. With the exception of Proteon's ProNET LANs, most of these LANs provided from less than 1 Mbps to 2.5 Mbps. LocalTalk provides just 233 kbps over UTP telephone wire, but it was originally intended primarily for printer sharing. Arcnet provided 2.5 Mbps over both coax and UTP. Proteon's LANs supported data rates of 10 Mbps and 80 Mbps, were both based on token-passing ring technology, and predate the 802.5 token ring standards by several years.

Table 2.2 summarizes the history of token ring.

By the end of the 1980s Ethernet held at least 50% of the total LAN market. Today Ethernet still dominates the worldwide LAN market, commanding well over 90% of all new LAN sales. Any LAN technology that was not an IEEE or ANSI standard was doomed. All other LAN technologies have fallen into relative obscurity, including several IEEE and ANSI standard LANs.

For all their supposed benefits over Ethernet, token ring and FDDI/CDDI could not overcome the forces of economics. While the later implementations of each relied on the same cabling technology, the cost of token ring and CDDI hardware remained outrageously high. And although the potential fault-tolerant benefits of FDDI were real, it was difficult for most businesses to justify the added expense of implementing a fully fault-tolerant topology, thereby eliminating one of the key benefits of FDDI. The only benefit to FDDI that remained was its higher speed. If an organization required a 100 Mbps LAN in 1987, FDDI was the only solution.

In its heyday token ring enjoyed over a third of the LAN market, but the introduction of 10Base-T Ethernet in 1987 curbed the expansion of token ring's market share. Prior to the appearance of 10Base-T Ethernet, which ran over ordinary UTP DIW-24 telephone wire, token ring installations required STP. STP is heavier, bulkier, more difficult to install and terminate, requires proper grounding, and is

TABLE 2.2 THE HISTORY OF TOKEN RING

Year	*Organization/Company*	*Standard/Product*	*Description*
1981	Proteon	ProNET/10	10 Mbps token passing ring
1983	Proteon	ProNET/80	80 Mbps token passing ring
1985	IBM/IEEE	802.5	4 Mbps token passing ring (TRN)
1987	ANSI	X3T12	100 Mbps token passing ring (FDDI)
1988	IBM/IEEE	802.5	16 Mbps token passing ring (TRN)
1998	IBM/IEEE	802.5	100 Mbps token passing ring (HSTRN)

much more expensive than UTP. Eventually UTP support was added to token ring and the speed was increased from 4 Mbps to 16 Mbps, and eventually to 100 Mbps. However, by this time 10Base-T Ethernet was well on the way to dominance.

The purported technological benefits of token ring were always dubious at best. Token ring has no mechanism to provide the kind of fault tolerance that FDDI can, so on that point token ring and Ethernet are equal, but token ring has always cost at least five times as much as Ethernet.

Marketed as having a "deterministic" network access method, the implication was that token ring LANs could provide controlled Quality of Service (QoS). Contrary to common belief, the reality of the technology was quite different. The token-passing network access method only ensures that as network traffic increases the LAN performance will deteriorate equally for all stations. It is an egalitarian network access method. This is not the same thing as providing a constant level of throughput or a guaranteed Quality of Service.

Token ring does support a priority/reservation system whereby certain stations may be granted more access to the ring than others; however, until recently there have been no operating systems or applications that take advantage of this feature. Furthermore, if the priority system were implemented it would obviate token ring's deterministic behavior! On this point, token ring was ahead of its time, but the point is moot. The commonly accepted upgrade path for 4 or 16 Mbps token ring is 100 Mbps Fast Ethernet and/or 1000 Mbps Gigabit Ethernet.

During the initial development of 100 Mbps Fast Ethernet, several approaches were submitted for consideration. The IEEE 802.3 committee submitted proposals for a two-pair solution and a four-pair solution, in addition to an optical fiber solution. These proposals are now known as IEEE 802.3u, which encompasses the following standards: 100Base-TX, 100Base-T4, and 100Base-FX.

Meanwhile, Hewlett-Packard proposed its own 100 Mbps solution called 100VG-AnyLAN. This network was designed to support 100 Mbps over Voice Grade UTP, and provided support for Ethernet or token ring frames (but not both at once). Initially 100VG-AnyLAN was supposed to support four pairs of UTP cabling with plans to add support for two pairs of UTP in the future. However, 100VG-AnyLAN did not support the CSMA/CD network access protocol and was therefore excluded from consideration as a potential Fast Ethernet standard by the 802.3u committee. Instead, 100VG-AnyLAN used a unique network access protocol called Demand Priority Access Method (DPAM). Because of this, the IEEE established a separate committee just for 100VG-AnyLAN named 802.12. Although the technology appeared to have broad support during its developmental stage, IEEE 802.12/100VG-AnyLAN never achieved any market penetration and today is just an historic footnote.

Although 10 Mbps Ethernet technology is now considered a *legacy LAN*, many of the small office, home office (SOHO) internetworking devices that are available still provide only a 10Base-T interface. It is important to note that the original implementations of Ethernet that relied on thick or thin coaxial cabling have been obsolete for some time. Relegated to operating at just 10 Mbps, the only use for coaxial cable in modern LANs is as a pull-rope to assist in the installation of new UTP and optical fiber cable plants.

Today's dominant LAN technologies are 100 Mbps Fast Ethernet and 1000 Mbps Gigabit Ethernet. The dominant LAN media are Cat5/Cat5E UTP and multimode optical fiber. Other wired network technologies such as Asynchronous Transfer Mode (ATM) over copper (155 Mbps) have also experienced a modicum of success. Other networks experiencing explosive growth are wireless LAN technologies such as IEEE 802.11b "WiFi" (Wireless Fidelity), which supports data rates from 1 Mbps to 11 Mbps. Also gaining in popularity are IEEE 802.11a and 802.11g, which support data rates up to 54 Mbps. 802.11g is backward compatible with 802.11b.

TRANSMISSION METHODS: BASEBAND AND BROADBAND

The transmission method used by virtually all LANs is baseband. Baseband transmissions are digital. Baseband is less costly and less complex to both implement and maintain. Baseband coax Ethernet LANs require only a NIC, a transceiver (which is typically built into the NIC), and the common medium to make a network. Hub- and switch-based technologies such as 10Base-T and 100Base-T Ethernet are also baseband LANs.

Networks that use a broadband transmission method differ in several ways. In this context, broadband transmissions are analog. Broadband transmissions require the use of two channels for data: one to transmit and one to receive. All devices transmit on one channel and receive on the other. Broadband data networks require a modem (which is typically a stand-alone unit, but may be a card installed in a computer), and the common medium is always coaxial. A device called a *head end*, or translator, removes transmitted data from the transmit channel, changes its frequency, and retransmits the data onto the receive channel.

Most baseband networks support symmetrical transmissions, whereas many broadband networks are asymmetrical. In other words, baseband systems transmit and receive at the same data rate and broadband systems often transmit and receive at different data rates. A popular example of asymmetrical broadband networking is cable modem Internet access. These metropolitan area networks (MANs) commonly support upstream (upload/transmit) data rates of 128 kbps or 256 kbps, and downstream (download/receive) data rates of 1.5 Mbps to several megabits per second. Another example is the so-called "56k" V.90 modem standard that can provide downstream data rates as high as 53 kbps but upstream data rates of just 33.6 kbps.

Some older broadband networks used the IEEE 802.3 10Broad36 standard, which required each station to have its own modem to access a broadband network. These now-obsolete broadband modem networks supported data connections such as RS-232 at several kilobits per second or 802.3/Ethernet at 10 megabits per second. Later broadband Ethernet LANs required a NIC with an embedded transceiver and modem, a connection to the media, a head end unit, and a Ph.D. to design and maintain them. The vast majority of these early broadband networks generally supported only one application (such as data) and did not exploit the *multimedia* potential of the system. The cable television systems used in many hotels are examples of broadband networks, some of which support in-room account review and checkout, games, and Internet access in addition to the usual television programming.

Over the years the original definition of the term *broadband* became so distorted that the word evolved to refer to anything that involved high speed and wide bandwidth, including LAN, MAN, and WAN technologies such as FDDI, SONET, ISDN, and ATM. To compound the confusion, the term is today used to describe a variety of both digital and analog networks.

True broadband LANs are no longer being developed or sold, and all LANs and WANs, as well as most MANs, are digital. The one exception is cable television

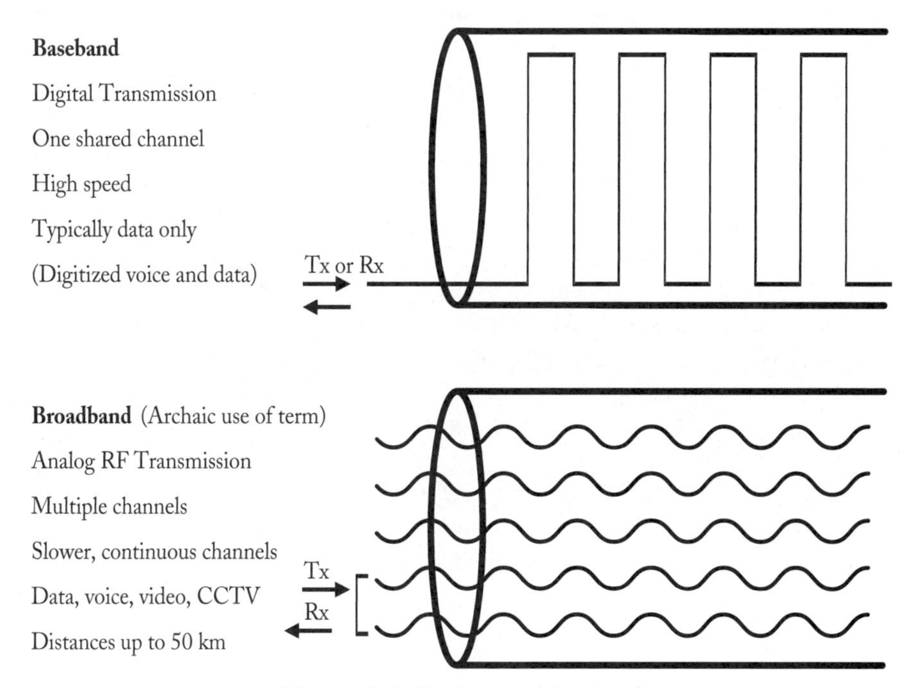

Figure 2.4 *Baseband and Broadband*
Digital baseband uses the entire bandwidth for each transmission whereas analog broadband bandwidth is divided into multiple channels.

networks, all of which are analog-based, most of which adhere to the IEEE 802.14 DOCSIS standard.

In contrast to digital baseband networks in which each transmission uses all available bandwidth, the bandwidth of analog broadband networks is divided into multiple channels. Each channel supports a separate transmission stream, allowing multiple concurrent transmissions, each using just a fraction of the total capacity of the media (see Figure 2.4).

CHAPTER SUMMARY

The path traveled by the LAN industry has not always been smooth or clearly marked, and quite a few wrong turns were taken along the way. Some wrong turns proved to be more difficult and expensive than others to correct, requiring that some hard decisions be made when the time came to upgrade or switch technologies. Ethernet is one of the oldest LAN technologies, and yet it has prevailed where newer technologies have failed. Arcnet succeeded because it was cheap, reliable, easy to install, widely supported, and cheap. Once the price of Ethernet fell to within range of Arcnet, Arcnet died. Along the way other LAN technologies were spawned but never gained adequate market share to sustain them. Token ring succeeded because of several factors: IBM promoted it, it employed a structured cable plant, it offered advanced features (albeit features that were rarely if ever used), IBM promoted it, and marketing touted its deterministic access method. Did I mention that IBM promoted it? Once 10Base-T Ethernet became available, token ring started losing ground. 10Base-T also supports a structured cable plant, and it uses UTP, which is less expensive and easier to install than token ring's STP. Even when it was new, 10Base-T Ethernet products cost a fraction of the price of token ring—typically about 20%. In spite of its fault-tolerant features, FDDI never gained much market share because of its high cost. The eventual development of 100Base-T Fast Ethernet was the death knell for both token ring and FDDI. The only remaining contenders were 100VG-AnyLAN and 100 Mbps High-Speed Token Ring, but they, too, fell by the wayside along with the other roadkill littering the paths to the *Information Superhighway*. Today 1000 Mbps Gigabit Ethernet is gaining popularity while 10 Gbps Ethernet and beyond are in the works.

REVIEW QUESTIONS

1. Ethernet was first developed by what organization?
2. Which LAN is most popular?
3. Which IEEE standard committee is responsible for Ethernet?
4. Which IEEE subcommittees defined Fast Ethernet?
5. Name three other LAN technologies.
6. Describe why token ring succeeded.
7. Describe why token ring failed.
8. Describe why Ethernet prevailed.
9. Define the following acronyms:

PARC	CAU
LAM	TRN
QoS	MAN
VG	FDDI
WiFi	Mbps

10. Compare baseband and broadband transmissions.

Standards and Organizations

OBJECTIVES

After completing this chapter, you should be able to:

- Define the various types of standards.
- Identify which technologies are standards and which are not.
- Describe the origins of Ethernet's standardization.
- Identify which standards organizations are responsible for LANs.
- Identify which standards organizations are responsible for cabling.
- Define the layers of the OSI model.
- Identify the relationship of various protocols to the OSI model.
- Explain why standardization is beneficial.

INTRODUCTION

Chapter 3 explains the quagmire we call "standards"—what they are and what they are not, who makes them and who breaks them. You will learn that while formal standardization has helped focus the LAN industry, stimulate competition, simplify integration, and reduce prices, the standardization process has also resulted in a few gaffes.

STANDARDS OVERVIEW

Several types of standards exist within the field of computer networking. There are industry standards and *de facto* standards, sometimes called *du jour* standards (or standards of the day). Industry standards are those that have been developed or refined within recognized standards-setting organizations and ultimately ratified as a standard. Some standards organizations, such as the American National Standards Institute (ANSI) and the National Electrical Manufacturers Association (NEMA), are, as the names imply, national organizations. Others, such as the International Standards

Institute and the International Telecommunication Union (ITU), are recognized internationally. De facto or du jour standards may be proprietary specifications or public domain specifications that have attained a high degree of popularity through widespread use. Two such examples would be IBM's Systems Network Architecture (SNA) protocol and Novell's IPX protocol. Neither are formal industry standards, but both have enjoyed broad support from numerous third-party vendors for nearly two decades.

In the early days of local area networking there were no standards. Various computer vendors addressed the networking needs of their customers by developing their own, proprietary solutions, such as Apple Computer Corporation's AppleTalk, Digital Equipment Corporation's DECnet, IBM's SNA, Novell's IPX, Wang's WangNet. Few of these proprietary networks are still sold today, but because of their widespread use and third-party support, IBM SNA and Novell IPX became de facto standards.

The recognition of similar networking requirements for personal computers prompted a plethora of products from a variety of vendors. These early efforts were proprietary, limited to data rates of several kilobits per second to a couple of megabits per second, and most have since been discontinued or replaced. However, the Xerox Corporation developed a LAN technology in the late 1970s that is still with us today. That LAN is Ethernet. Developed at the Xerox Palo Alto Research Center (Xerox PARC) and based on a network access protocol developed by the University of Hawaii called *Aloha*, Ethernet has evolved to support many types of media and data rates from 1 Mbps to 10,000 Mbps (10 Gbps). Credit for Ethernet's invention is awarded to Bob Metcalfe (see Figure 3.1).

The incredible success of Ethernet is due to several factors, probably the foremost being the cooperation between Xerox, Intel, and Digital Equipment Corporation (DEC) that resulted in the DIX Ethernet specifications. The earliest Ethernet specifications

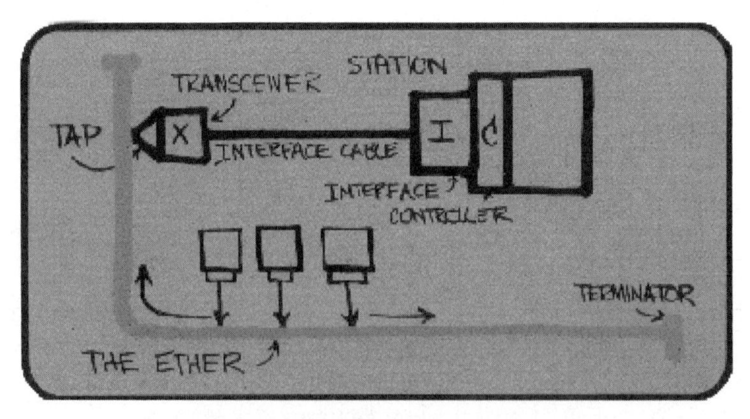

Figure 3.1 *Original Ethernet Concept*
Bob Metcalfe's original conceptual Ethernet drawing.

were called Ethernet Specification 1 and 2, or ESPEC1 and ESPEC2, also known as Bluebook-1 and Bluebook-2 (named for color of the book cover). The DIX team submitted their work to two standards organizations for consideration; they were ANSI and the Institute of Electrical and Electronic Engineers (IEEE). In February of 1980 the ANSI/IEEE 802 Standard Committee was established to define local and metropolitan area networks (LANs and MANs). Throughout the 1980s the importance of network standards became increasingly apparent. As a result, virtually all manufacturers of networking products moved to embrace new standards as they emerged—especially the ANSI/IEEE LAN standards.

IEEE standards are developed within the Technical Committees of the IEEE and the Standards Coordinating Committees of the IEEE Standards Board. During the 1980s the IEEE collaborated with ANSI on numerous networking standards development efforts, but while working on the ANSI/IEEE 802.8 fiber network specifications the two bodies diverged. In the end the IEEE standard 802.8 high-speed optical fiber network lost to the ANSI standard X3T9, now known as FDDI (ANSI X3T12).

In 1985, ISO Technical Committee 97, Information Processing Systems, adopted IEEE standard 802.3-1985 as draft International Standard ISO/DIS 8802-3. The standard was subsequently approved by the International Organization for Standardization (ISO) in 1989, and various amendments to the standard were added within a few years of being approved by ANSI/IEEE. During the 1980s and early 1990s, the IEEE and ANSI worked cooperatively with the ISO, conforming their standards to fit the ISO's Open System Interconnection (OSI) seven-layer reference model. These efforts resulted in some winners and some losers—by today's count, mostly losers.

STANDARDS ORGANIZATIONS

EIA

The Electronic Industry Alliance, accredited by ANSI, provides a forum for industry to develop standards and publications in major technical areas: electronic components, consumer electronics, electronic information, and telecommunications. URL: www.eia.org

TIA

The Telecommunications Industry Association represents the communications sector of the EIA. URL: www.tiaonline.org

NEMA

For more than seventy years, the National Electrical Manufacturers Association has been developing standards for the electrical manufacturing industry and is today one of the leading standards development organizations in the world. As such, it contributes to an orderly marketplace and helps ensure the public safety. NEMA attempts to promote the competitiveness of its member companies by providing a forum for:

> The development of technical standards that are in the best interests of the industry and the users of its products; the establishment and advocacy of industry policies on legislative and regulatory matters that might affect the industry and those it serves; and the collection, analysis, and dissemination of industry data. (www.nema.org)

NEMA publishes over 200 standards and offers them for sale, along with certain standards originally developed by the ANSI and the International Electrotechnical Commission. The association promotes safety in the manufacture and use of electrical products, provides information about NEMA to the media and the public, and represents industry interests in new and developing technologies. URL: www.nema.org

IEC

The International Electrotechnical Commission, known as IECQ, prepares and publishes international standards for all electrical, electronic, and related technologies. These serve as a basis for national standardization and as references when drafting international tenders and contracts. The IECQ promotes international cooperation on all questions of electrotechnical standardization and related matters, such as the assessment of conformity to standards, in the fields of electricity, electronics, and related technologies. The IECQ charter includes all electrotechnologies including electronics, magnetics and electromagnetics, electroacoustics, multimedia, telecommunication, and energy production and distribution, as well as associated general disciplines such as terminology and symbols, electromagnetic compatibility, measurement and performance, dependability, design and development, safety and the environment. URL: www.iec.ch

ANSI

The American National Standards Institute is the official U.S. representative to the International Accreditation Forum (IAF), the ISO, and, via the U.S. National Committee, the International Electrotechnical Commission. ANSI ("ann-see") is also the U.S. member of the Pacific Area Standards Congress (PASC) and the Pan American Standards Commission (COPANT). ANSI is a private, nonprofit organization that administers and coordinates the U.S. voluntary standardization and conformity assessment system. The Institute's mission is to enhance both the global competitiveness of U.S. business and the U.S. quality of life by promoting and facilitating voluntary consensus standards and conformity assessment systems, and safeguarding their integrity. URL: www.ansi.org

IEEE

The Institute of Electrical and Electronic Engineers helps advance global prosperity by promoting the engineering process of creating, developing, integrating, sharing, and applying knowledge about electrical and information technologies and sciences for the benefit of humanity and the profession. URL: www.ieee.org

ISO

The International Organization for Standardization is a worldwide federation of national standards bodies from some 140 countries, one from each country. ISO is a nongovernmental organization established in 1947. The mission of ISO is to promote the development of standardization and related activities in the world with a view to facilitating the international exchange of goods and services, and to developing cooperation in the spheres of intellectual, scientific, technological, and economic activity. ISO's work results in international agreements, which are published as International Standards. URL: www.iso.org

ITU-T

The ITU-T Telecommunication Standardization Sector (ITU-T) is one of the three Sectors of the ITU. ITU-T was created on March 1, 1993, within the framework of the "new" ITU, replacing the former International Telegraph and Telephone Consultative Committee (CCITT), whose origins go back to 1865. The ITU-T mission is to ensure an efficient and on-time production of high-quality standards covering all fields of telecommunications except radio aspects.

Standardization work is carried out by 14 study groups in which representatives of the ITU-T membership develop recommendations for the various fields of international telecommunications on the basis of the study of questions (i.e., areas for study). At present, more than 2600 recommendations (standards) on some 60,000 pages are in force. URL: www.itu.int (specifically, www.itu.int/ITU-T/).

IETF

The Internet Engineering Task Force is a large open international community of network designers, operators, vendors, and researchers concerned with the evolution of the Internet architecture and the smooth operation of the Internet. It is open to any interested individual. The actual technical work of the IETF is done in its working groups, which are organized by topic into several areas (e.g., routing, transport, security). Much of the work is handled via mailing lists. The IETF holds meetings three times per year.

The IETF working groups are grouped into areas, and managed by Area Directors, or ADs. The ADs are members of the Internet Engineering Steering Group (IESG). Providing architectural oversight is the Internet Architecture Board, (IAB). The IAB also adjudicates appeals when someone complains that the IESG has failed. The IAB and IESG are chartered by the Internet Society (ISOC) for these purposes. The general area director also serves as the chair of the IESG and of the IETF, and is an ex-officio member of the IAB.

The Internet Assigned Numbers Authority (IANA) is the central coordinator for the assignment of unique parameter values for Internet protocols. The IANA is chartered by the ISOC to act as the clearinghouse to assign and coordinate the use of numerous Internet protocol parameters. URL: www.ietf.org.

ISO OSI SEVEN-LAYER REFERENCE MODEL

The ISO Open Systems Interconnection seven-layer reference model was developed to provide a framework within which network communications may be defined. Each of the seven layers is responsible for a specific range of tasks required to provide various types of network communications. Some network applications require guaranteed data delivery whereas others do not. There are also situations where guaranteed delivery would be detrimental to an application's performance. To satisfy these diverse communications requirements different protocols that correspond to various layers of the OSI model have been adopted or developed from scratch. Today this model should be considered prescriptive rather than descriptive. In other words, the

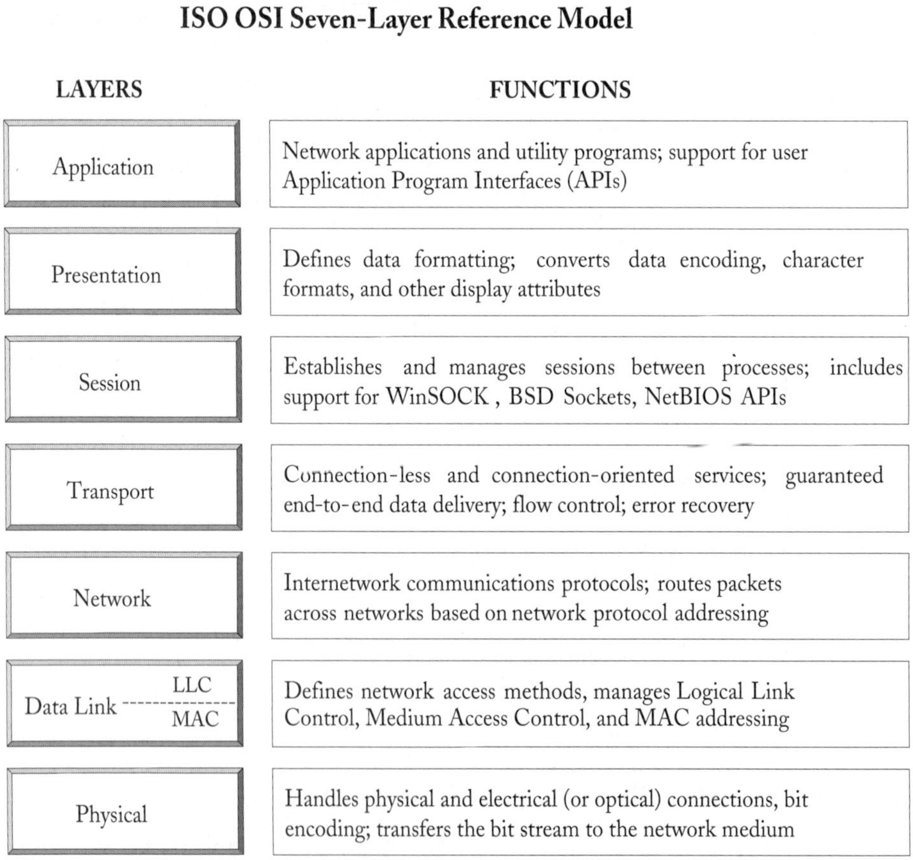

ISO OSI Seven-Layer Reference Model

LAYERS	FUNCTIONS
Application	Network applications and utility programs; support for user Application Program Interfaces (APIs)
Presentation	Defines data formatting; converts data encoding, character formats, and other display attributes
Session	Establishes and manages sessions between processes; includes support for WinSOCK , BSD Sockets, NetBIOS APIs
Transport	Connection-less and connection-oriented services; guaranteed end-to-end data delivery; flow control; error recovery
Network	Internetwork communications protocols; routes packets across networks based on network protocol addressing
Data Link (LLC / MAC)	Defines network access methods, manages Logical Link Control, Medium Access Control, and MAC addressing
Physical	Handles physical and electrical (or optical) connections, bit encoding; transfers the bit stream to the network medium

Figure 3.2 *ISO OSI Seven-Layer Reference Model*
Functional descriptions of each layer of the ISO OSI Seven-Layer Reference Model.

model is used to explain how network communications do function, not dictate how they must function. The seven layers of the OSI model and their functions may be seen in Figure 3.2.

ISO AND PROTOCOL COMPARISON

Virtually all protocols, proprietary and standard, are based on architecture of a layered model. Network communications have long been defined in a hierarchy of functions that requires the cooperation of an entire suite of protocols. This is true for IBM's SNA, DEC's Digital Network Architecture (DNA), the ISO OSI, and the IETF's Transmission Control Protocol/Internet Protocol (TCP/IP) (see Figure 3.3).

IEEE protocols are based on a three-layer architecture that corresponds to the bottom two layers of the OSI model. The IEEE Logical Link Control (LLC), Media Access Control (MAC), and Physical layers correspond to the ISO OSI Data Link and Physical layers. For this reason, the OSI Data Link layer is usually illustrated as having two sub-layers—LLC and MAC.

The original DIX ESPEC Ethernet defines functions that correspond to the two bottom layers of the OSI model—the Data Link and Physical layers. When the IEEE adopted DIX Ethernet they made several changes to the specification, one of which

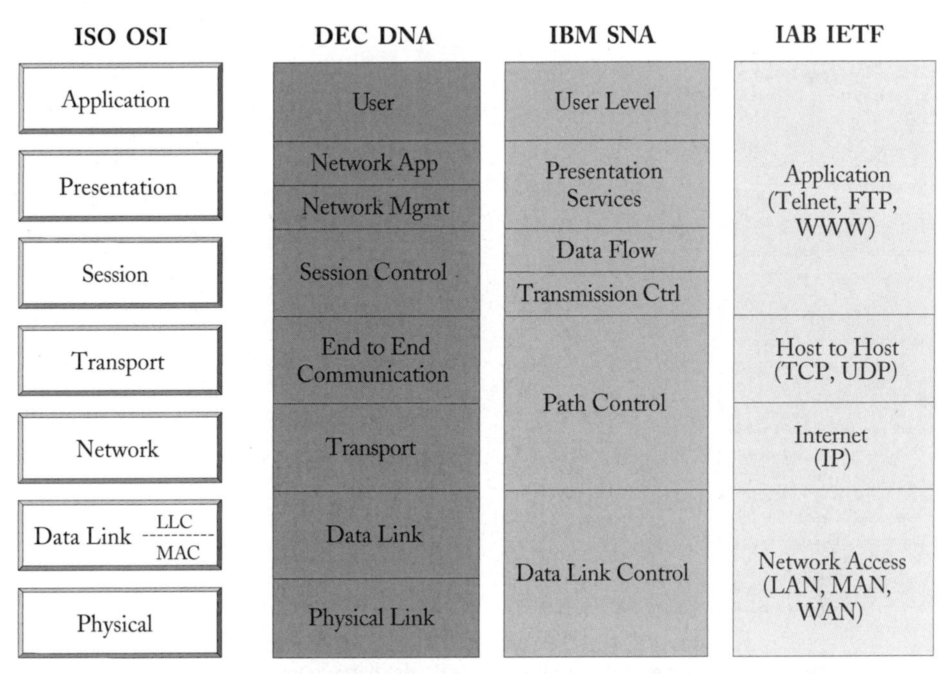

Figure 3.3 *Protocol Stack Comparison*
The OSI model provides an international standard to which all other protocols may be compared.

was to alter the protocol to comply with the IEEE three-layer model, and thereby adhere to the OSI seven-layer model. This involved changing the function of the 2-byte Ethernet Type field into a 2-byte Length field. The protocol was also restructured to comply with the bottom one-and-a-half layers of the OSI model—the Physical layer and the MAC sub-layer (see Figure 3.4).

The 2-byte Ethernet Type (or EtherType) field is sometimes referred to as the "Protocol Type" field because it defines the upper layer protocol contained within the Ethernet frame's Data field. This 2-byte field contains a number used to identify a given protocol and is represented in hexadecimal. Examples of such upper layer protocol numbers include 0800h (IP) and 0806h (ARP). The IEEE replaced the functionality of the Ethernet Protocol Type field with two fields in the IEEE 802.2 Logical Link Control (LLC) protocol. However, instead of using the established 2-byte protocol identification numbers used in DIX Ethernet, the IEEE developed its own 1-byte numbering scheme.

One can only suppose that the standards groups expected protocol engineers everywhere to rewrite their network communications protocols to work with the new IEEE 802.3 implementation of Ethernet or drop them entirely to adopt the new ISO

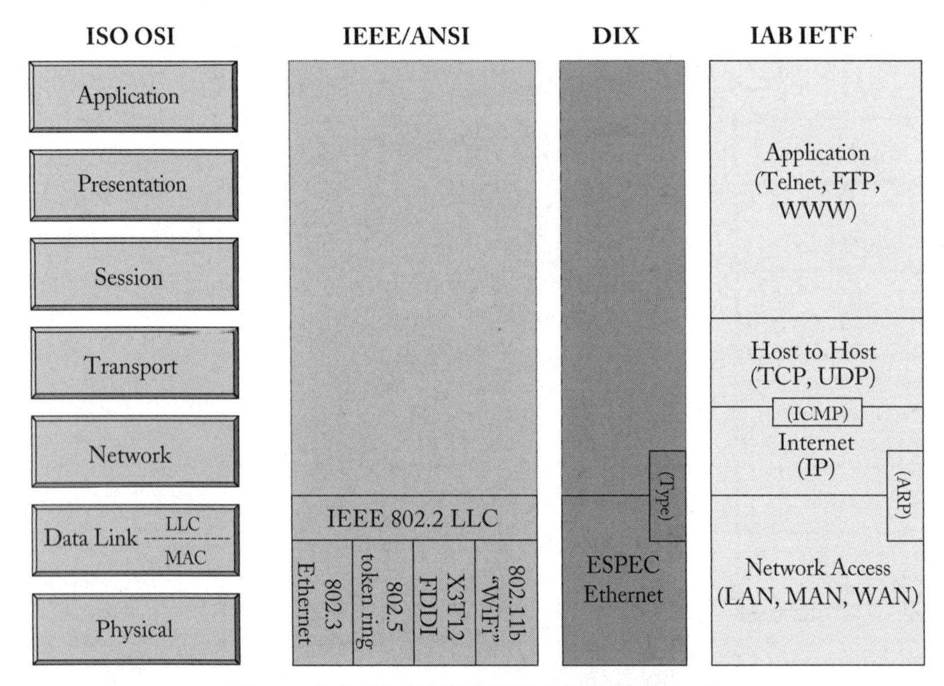

Figure 3.4 *ISO OSI, IEEE/ANSI, DIX, IAB IETF stacks*
The IEEE and ANSI developed network standards that adhere to the OSI model; however, DIX and IETF protocols predate these standards.

OSI suite of protocols. Certainly this is what the ISO had in mind as they developed the Open System Interconnection protocol suite. While a few vendors such as Apple and Novell did in fact modify their protocols to accommodate the new IEEE standard LANs, most just held their ground with DIX Ethernet.

The IETF addressed the problem by developing a new protocol extension to IEEE 802.2 LLC. Referred to as the Sub-Network Access Protocol (SNAP), or 802.2 SNAP, this extension replaces the Ethernet Protocol Type field. The modification allows IEEE standard LANs to support other non-OSI protocols that rely on the information contained within the DIX Ethernet Protocol Type field.

Before the release of IEEE 802.3 Ethernet, Apple had supported its AppleTalk protocol over the proprietary LocalTalk LAN and DIX Ethernet. AppleTalk over Ethernet is known as EtherTalk. The configuration of AppleTalk over DIX Ethernet is now referred to as AppleTalk Phase 1. AppleTalk Phase 2 was developed to support the AppleTalk protocol using IEEE 802.3 standard Ethernet frames.

Novell embraced IEEE standards early on, even before the IEEE 802.2 LLC standard was completed. To support the Internetwork Packet eXchange (IPX) protocol over IEEE 802.3 Ethernet, Novell hard-coded the checksum field of the IPX packet header (the first 16 bits) to all binary ones. This established a unique method for identifying IPX packets contained in IEEE 802.3 standard frames and is commonly referred to as "802.3 *raw*." Novell later modified IPX to provide full support for 802.2 LLC, but to maintain backward compatibility support was continued for the 802.3 raw frame format. Today Novell provides native support for the IP protocol and DIX Ethernet frames, as well as their legacy IPX protocol and various LAN frame options.

Networking devices such as repeaters, bridges, routers, and gateways generally correspond to the functional descriptions of the OSI model. Some devices such as "switches" may correspond to the defined functionality of more than one OSI layer. Some network switches perform their functions based on layer 2 information (bridging), some perform their functions based on layer 3 information (routing), while other switches rely on layer 4 information (see Figure 3.5). The functionality of certain protocols, such as the Address Resolution Protocol (ARP), tends to overlap OSI layers. The role of ARP is to resolve Data Link (MAC) layer (or hardware) addresses to Network layer addresses.

SO WHY STANDARDIZE?

We have seen that standardization has not always resulted in winning solutions, and in fact has caused some problems for existing technologies. Still, standardization is good for consumers and businesses because it provides a fertile foundation for competition to flourish. Competition drives prices down, expands the target market, and provides for greater technological diversity than would likely arise from the closed environment of proprietary products. This has resulted in a broad range of options, rapid

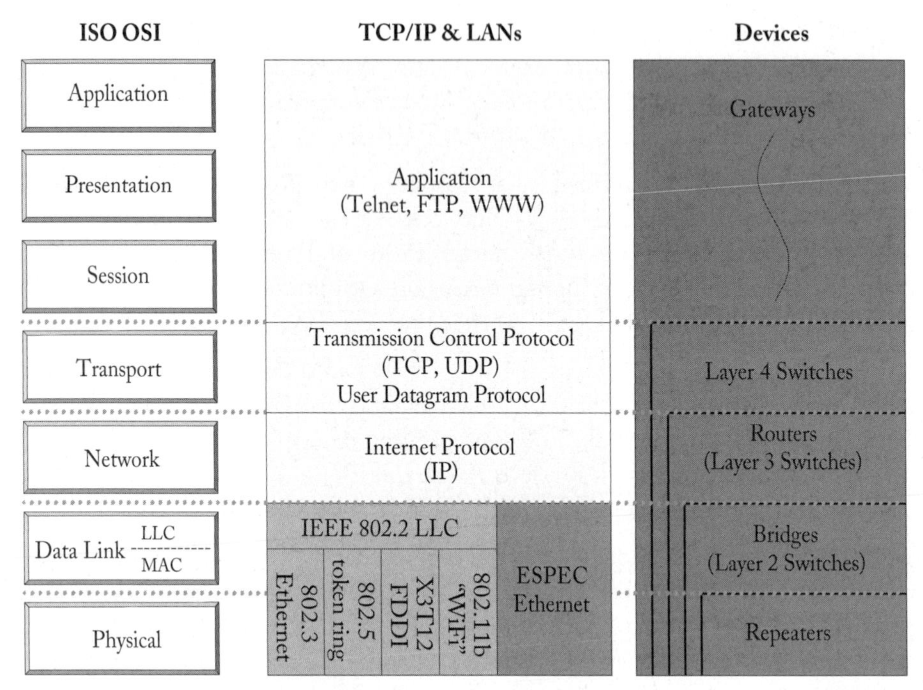

Figure 3.5 *ISO OSI, IEEE, and Network Devices*
All standard network technologies support virtually all protocol suites such as TCP/IP.

acceptance of new technologies and rapid advancement of still newer technologies—as production levels increase, so does reliability.

During the 1980s this rampant race toward ever more standardization drove the standards organizations to embrace yet other standards. In the effort to standardize Ethernet and comply with the ISO OSI standards model, ANSI/IEEE 802.3 essentially produced a dysfunctional version of Ethernet. The IETF salvaged the situation by developing SNAP. Until the mid-1990s most networking standards bodies really thought that the ISO OSI protocol suite would eventually displace all other networking protocols. Then the World Wide Web happened, the Internet experienced explosive growth, and the Internet protocols (TCP/IP) became the global communications standard that everyone thought would be the destiny of the ISO OSI.

Toward the end of the 1990s the LAN market started to thin out. FDDI and CDDI remained too expensive for most businesses to even consider, token ring was already losing ground to 10Base-T Ethernet, and when 100Base-T Fast Ethernet appeared it was all over. 100 Mbps Fast Ethernet was a fraction of the price of 4/16 Mbps token ring, and a tiny fraction of the price of 100 Mbps FDDI/CDDI. By this time, LAN switches were replacing hubs, thereby eliminating Ethernet's channel contention issues and securing its position as the dominant LAN on the planet. Fast Ethernet

NICs can now be purchased for under $10 US (apiece), and small (unmanaged) 10/100 Mbps Ethernet switches cost less than $10 per port. The drive is now on to move to Gigabit Ethernet (1000 Mbps) and 10 Gigabit Ethernet (10,000 Mbps).

Economics "culls the herd." By weeding out unviable technologies, it advances superior technologies and makes room for fresh innovation. Standardization provides a structured framework that supports innovation by leveraging established technologies. Innovation outside of the standards is what we call proprietary. This is not to say that independent organizations cannot create some excellent innovations without the benefit of standardization, but it is certainly more difficult, and in the computer networking industry it is quite rare. Many successful technologies such as Ethernet were in fact invented inside individual corporations, but were later submitted to various standards bodies for further review and eventually ratification as standards. So in conclusion, although standards groups have committed their fair share of gaffes, overall their work has been very beneficial to us all.

CHAPTER SUMMARY

The LAN market consists of a wide array of technologies, some formal standards, some not so formal; still others remain proprietary. There are many standards organizations involved in the many facets that make up the LAN industry, and they do not all cooperate. As strange as it may sound, the standards organizations responsible for furthering LAN technologies such as Gigabit Ethernet are not the same standards organizations responsible for defining the media that will support it! Case in point: The TIA/EIA defined Category 5 UTP with the intent that it should support future applications such as Gigabit Ethernet; however, once the two technologies were mated a few problems emerged. These problems were addressed by the release of a new and improved *Category 5E* (Enhanced) UTP specification that supercedes the original Category 5 standard. They call it the *bleeding edge* for a reason.

REVIEW QUESTIONS

1. Define the various types of standards.
2. List three proprietary LAN technologies.
3. List two standards organizations responsible for LAN technologies.
4. List four standards organizations responsible for cabling technologies.
5. Identify the layers of the OSI model.

6. Which OSI layer contains the LLC and MAC sub-layers?
7. LAN technologies are defined in which layers of the OSI model?
8. Identify the relationship of various protocols to the OSI model.
9. Explain why standardization is beneficial.
10. Define the following acronyms:

ANSI	NEMA
ISO	ITU
SNA	IEC
IETF	OSI
TCP	IP
LLC	MAC

Data Elements

OBJECTIVES

After completing this chapter, you should be able to:

- Define bits and bytes and the relationship between them.
- Convert binary values to their decimal equivalent.
- Understand the meanings of kilo-, mega-, giga-, and tera-.
- Define the difference between frames, packets, and cells.
- Identify the data elements that correspond to each OSI layer.
- Define MAC address and OUI.
- Describe the relationship between segments, packets, and frames.
- Explain the difference between an OS and an NOS.
- Identify common operating systems and network operating systems.

INTRODUCTION

Chapter 4 literally gets into the bits and bytes of data transmissions. Binary is the fundamental building block of all data processing and data communications. To be stored and manipulated by computers, all types of information either originates as binary data or is converted from an analog form into binary data. These basic elements are grouped into chunks to form bytes, which are grouped into even larger chunks to form meaningful information. Each step in the data communications process manipulates different arrangements of these chunks.

BITS, BYTES, FRAMES, PACKETS, AND CELLS

The smallest unit of data is a *bit*—a binary number with a value of 0 or 1. Multiple bits can be grouped to represent larger values. A group of seven or eight bits is called a *byte*, while a group of four bits is called a *nibble*. A byte may represent an alphanumeric character (A, B, C, 1, 2, 3) or a control code, or it may be part of a program. In addition to the computer programs, all data documents, images, and music are represented digitally using groups of bits and bytes. All of the bytes representing a given word processing document are grouped to form a *file*. Files may be stored on disk drives, loaded into computer memory, or transmitted to another computer over a network.

Networks rely on still more groups of bytes used to structure a bit stream for transmission over the network media. Networks such as Ethernet and token ring define groups of bits and bytes called *frames*. Networks also rely on additional communications protocols such as TCP/IP or Novell's IPX. These communications protocols define groups of bits and bytes called *packets*. The information contained in frames and packets may consist of just a few bits or several 8-bit bytes. The user's data, such as a word processing document, is the payload and is encapsulated within the packets of the communications protocols, which are in turn encapsulated within network frames and transmitted across the network as a stream of bits.

BITS AND BYTES

The elements that make up network transmissions are bits, bytes, frames, packets, and/or cells. As mentioned above, a bit is the most basic unit of data. A single bit may have a value of either zero or one. Mathematics refers to this as *binary* or *Base2*. We humans tend to prefer *decimal* or *Base10*. This preference is attributed to the fact that human biology provides us with a total of ten digits—five on each hand. While humans have adapted rather well to thinking in Base10, most computers are still functioning on Base2. Base2 lends itself to computing because computers still rely on electricity and on switches that close and open circuits—to turn the flow of that electricity on and off. In most systems, a bit value of one (1) represents "on" while a bit value of zero (0) represents "off." The central processing unit (CPU) in a computer is actually made up of millions of tiny on/off switches, and the memory and disk drives in a computer store data as bits. All forms of data are passed back and forth between the disk drive, CPU, memory, and peripheral devices as streams of bits. Data is also transmitted over networks between computers as streams of bits. When data is transmitted over the internal bus of a computer, many bits (8, 16, 32, or 64) are transmitted simultaneously in parallel. When data is transmitted over most networks, single bits are transmitted serially, one at a time. There are, of course, exceptions. Common examples would be the parallel printer cable that connects a printer to a computer LPT port, or computer disk-drive interfaces (such as IDE and SCSI) that consist of a flat, multi-conductor cable.

To represent values larger than just zero or one, multiple bits may be grouped together. A group of eight bits constitutes a byte. Each bit position within a byte represents a different value. The rightmost bit typically represents the lowest value and the leftmost

Bit Values	128	64	32	16	8	4	2	1
Bit Positions	8	7	6	5	4	3	2	1
	MSB							**LSB**

Bit #1 represents a decimal value of zero or 1 (LSB, the least significant bit)
Bit #2 represents a decimal value of zero or 2
Bit #3 represents a decimal value of zero or 4
Bit #4 represents a decimal value of zero or 8
Bit #5 represents a decimal value of zero or 16
Bit #6 represents a decimal value of zero or 32
Bit #7 represents a decimal value of zero or 64
Bit #8 represents a decimal value of zero or 128 (MSB, the most significant bit)

bit represents the highest value. These are referred to as the least significant bit and the most significant bit, respectively. To identify each *bit position*, number the bits 1 through 8 from right to left: *8, 7, 6, 5, 4, 3, 2, 1.* The actual value of each bit position increases exponentially from right to left. In other words, the bit value doubles from one bit position to the next: 128, 64, 32, 16, 8, 4, 2, 1. (See chart above.)

To count in binary, remember that a one means "on" and a zero means "off." To convert the binary number into its decimal value, simply add the bit values wherever a binary one appears. For example, the following byte, 00000110, contains just two bits in the "on" position. These are the second and third bit positions, which have bit values of 2 and 4 respectively. To determine the decimal value, simply add the bit values: 2+4=6. In the next example the following byte, 01100110, contains four bits in the "on" position. These are the second, third, sixth, and seventh bit positions, which have bit values of 2, 4, 32, and 64, respectively. Again, to determine the decimal value, simply add the bit values: 2+4+32+64=102. The highest possible value represented by an 8-bit byte is 255, which is achieved by setting all eight bits to one (11111111); the lowest possible value is 0, which is achieved by setting all eight bits to zero (00000000). (See chart on next page.)

To represent values larger than 255, multiple bytes may be grouped together to form 16-bit *words*, or 32-bit words, or 64-bit words. Computer operating systems, programs, and CPUs usually limit word size.

Oddly, the terms bits and bytes often get used interchangeably, as if the difference is inconsequential. This is due in large part to the way the computer and networking industries measure the capacities of different technologies. File sizes and disk-drive capacities are always stated in bytes, for example: kilobytes (kB), megabytes (MB), gigabytes (GB), or terabytes (TB). Network data rates are stated in bits over time, for example: kilobits per second (kbps), megabits per second (Mbps), gigabits per second

Bit Values	128	64	32	16	8	4	2	1
Bit Positions	8	7	6	5	4	3	2	1
MSB							**LSB**	

Decimal:	Binary:							
0 =	0	0	0	0	0	0	0	0
1 =	0	0	0	0	0	0	0	1
2 =	0	0	0	0	0	0	1	0
3 =	0	0	0	0	0	0	1	1
4 =	0	0	0	0	0	1	0	0
5 =	0	0	0	0	0	1	0	1
6 =	0	0	0	0	0	1	1	0
7 =	0	0	0	0	0	1	1	1
8 =	0	0	0	0	1	0	0	0
128 =	1	0	0	0	0	0	0	0
255 =	1	1	1	1	1	1	1	1

(Gbps), or terabits per second (Tbps). The lowercase "k" for kilo and uppercase "M-G-T" for mega-, giga-, and tera-, respectively, is technically correct.

The use of lowercase "b" for bits and uppercase "B" for bytes is a convention to which too few people adhere. Failing to clearly identify this little detail leaves it up to the reader to determine the correct meaning based solely on context. To confuse matters even more, the interface characteristics of disk-drive interface technologies such as SCSI and IDE are sometimes referenced in either bytes or bits. As if this was not bad enough, some disk-drive manufacturers use these terms inconsistently in the specifications of a single product. For example, they may define 1-kilobit to mean 1000 bits, but 1-kilobyte to mean 1024 bytes. While it is true that the pedestrian usage of the term kilo is understood to mean 1000, the computer usage of kilo is derived from binary multiples—2*2*2*2*2*2*2*2*2*2, or 2^{10}, which equals 1024. Fortunately, when talking about megabytes, gigabytes, and terabytes of disk storage and data rates, a few kilobytes are rarely missed. And in general, file sizes and disk-drive capacities are measured in bytes and network data rates are measured in bits per second.

Again, it is important to learn to use industry jargon and acronyms properly. Far too many people in this industry know just enough to be dangerous and spew techno-babble on a regular basis. Correct use of technical language can reinforce the perceived validity of one's position, whereas using technical language incorrectly can only serve to undermine one's credibility. If you cannot talk the talk few people will give you the opportunity to walk the walk.

FRAMES

To transmit data across networks, bytes are grouped into various fields that constitute *frames*. With few exceptions, all networks use frames. The structure of those frames varies from one network technology to another. Examples of networks that use frames are Ethernet, token ring, and FDDI. Each different network technology defines a unique frame structure—the frames used by Ethernet, token ring, and FDDI are not compatible with each other. One example of a network that does not use frames is ATM, which instead uses fixed length packets called cells.

Frames are defined by the networking standards that correspond to the ISO OSI Data Link layer (layer 2). As mentioned above, frames consist of fields that contain some number of bits or bytes. Frames typically contain addressing and control information in the header, and error-checking and sometimes control information in the trailer. As the names imply, a frame header is positioned at the beginning of each frame while the trailer is positioned at the end of each frame. Sandwiched in between the header and trailer is a data field—the header and trailer *frame* the data (see Figure 4.1).

All LAN frames consist of several basic fields, including a Destination Address, Source Address, one or more Control functions, upper-layer Data, and a Frame Check Sequence for error detection. ANSI/IEEE LANs use 6-byte addresses, often referred to as Media Access Control (MAC) addresses, "physical" or "hardware" addresses. The destination address identifies the recipient of the frame and the source

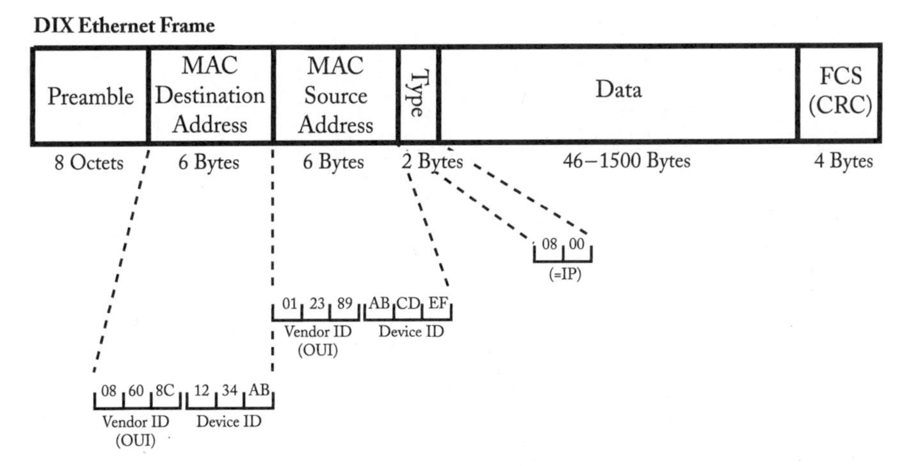

Figure 4.1 *Basic Ethernet Frame*
Details include destination and source addresses, and Ethernet Type field.

address identifies the sender. Source addresses are actually burned into chips on each network adapter or NIC and are unique to every product or interface manufactured.

All standard LANs support the same 6-byte addressing scheme that is maintained by the IEEE Registration Authority. The 6-byte address consists of a 3-byte prefix and a 3-byte suffix. To prevent duplicate addresses the IEEE established a Registration Authority to ensure that all manufacturers use a different prefix, or *Organizationally Unique Identifier* (OUI).

The term *Organizationally Unique Identifier* is long-winded standards language that simply means "Company ID." All manufacturers of network interface cards and other products that incorporate standards-based network technologies such as Ethernet and token ring must obtain their own unique 3-byte OUI from the IEEE Registration Authority.

The suffix, or Device ID, is unique to each network interface card or other network interface produced by that manufacturer. The combination of the OUI and Device ID provides each individual product with a globally unique MAC address. (It should be noted, however, that a few manufacturing errors have occurred wherein duplicate MAC addresses have in fact occurred.)

With a 3-byte Device ID each OUI can provide 16,777,214 unique MAC addresses. Even so, many vendors have more than one OUI. This is due to mergers, acquisitions, or because product sales have nearly consumed entire blocks of unique Device IDs. The complete 6-byte MAC address is expressed in hexadecimal and appears as follows: 00-00-1B-01-23-AF. More information about OUI and Company ID Assignments may be found here: http://standards.ieee.org/regauth/oui/index.shtml.

OUI Examples:

3Com	00-01-03	Cisco	00-00-0C
3Com	00-02-9C	Cisco	00-01-42
3Com	00-06-8C	Cisco	00-01-43
3Com	00-0A-04	Cisco	00-01-63
Intel	00-02-B3	SMC	00-04-E2
Intel	00-03-47	SMC	00-0B-C5
Intel	00-04-23	Novell	00-00-1B

PACKETS

The data field of most network frames contains upper layer protocol information and data. Some network standards refer to this as a protocol data unit (PDU) while other specifications simply call the combination a *packet*. With few exceptions, all network frames contain packets. One example of a network frame that does not always contain a packet is token ring. Token ring uses special frames called *tokens* that are used by the network access method to govern access to the media. Tokens are specialized 3-byte frames that provide some network access control information but contain no addressing or data. Token ring also uses special full-length frames to perform ring maintenance functions such as Active Monitor Present, Standby Monitor Present, Claim Token, and Beaconing. These functions rely only on the access control and frame control fields of the token ring frame to carry out their responsibilities.

The terms *packet* and *frame* are often used interchangeably, but they are technically not synonymous. Packets are defined by the network protocol specifications that correspond to the ISO OSI Network layer (layer 3). Like frames, packets consist of fields of some number of bits or bytes. Like frames, packets typically contain addressing and control information in the header, but unlike frames packets have no trailer. Packets consist only of the protocol header followed by a data field. The data field of a packet may contain additional upper layer protocols and/or application data. Examples of protocols that use packets are the Internet Protocol (IP) and Novell's Internetwork Packet eXchange (IPX).

In summary, application programs generate streams of data in bytes. Those byte streams are grouped into *blocks* of bytes, or *segments*, which are encapsulated in the data field of upper layer protocols such as TCP. TCP adds its own protocol header information and passes that TCP segment down the protocol stack to a Network layer protocol such as IP. IP encapsulates the TCP segment, adds its own protocol header information, and passes that IP packet down the stack to a Data Link layer protocol such as Ethernet. Ethernet then encapsulates the IP Packet in its data field, adds its own protocol header and trailer information, and passes that Ethernet frame down the stack to its Physical layer interface, which then encodes that frame as a bit stream for transmission onto the network medium (see Figure 4.2).

CELLS

There is always at least one exception to every rule and networking has plenty of them (both rules and exceptions). One of the most noteworthy is Asynchronous Transfer Mode. ATM is a network technology available for use in WANs as well as LANs. While ATM is rarely used in local area networks due to its cost and complexity, it has for years formed the core of most telecommunications carriers' nationwide infrastructures.

Rather than using frames to transmit data across networks, ATM uses fixed-length packets called *cells*. ATM cells are similar in structure to packets in that they include a header element and data field, but no trailer. Whereas packet technologies support variable-length packets, ATM defines a fixed-length cell consisting of a 5-byte header and 48-byte payload, for a total of 53 bytes (see Figure 4.3).

Like frames, cells correspond to the ISO OSI Data Link layer (layer 2), and consist of several fields that contain some number of bits or bytes. The addressing and control information in the header identifies a virtual circuit (path and channel), defines the payload type, sets priorities, provides a checksum for the header itself, and can define flow control—all in just 5 bytes.

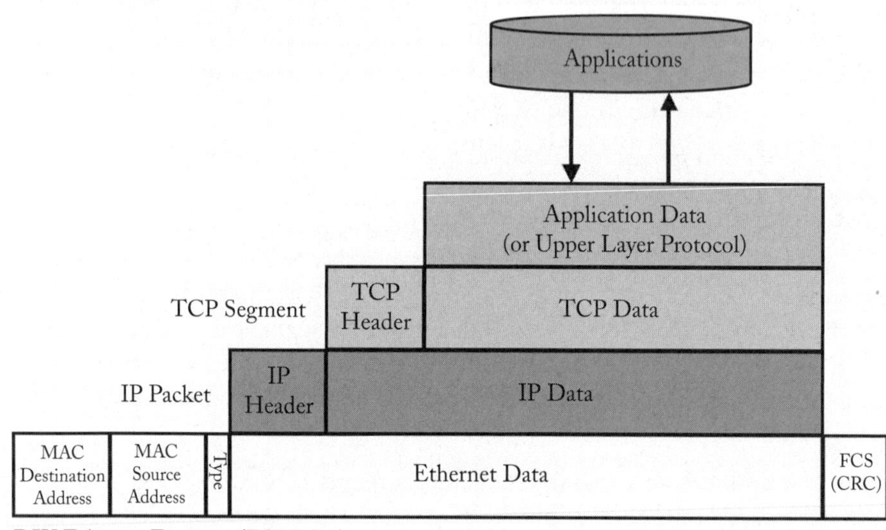

Figure 4.2 *Segment, Packet, Frame, and Bit Stream*
Applications send data as streams of bytes down the protocol stack where it is encapsulated in segments, packets, and then frames, before being encoded for transmission over the network media as a bit stream.

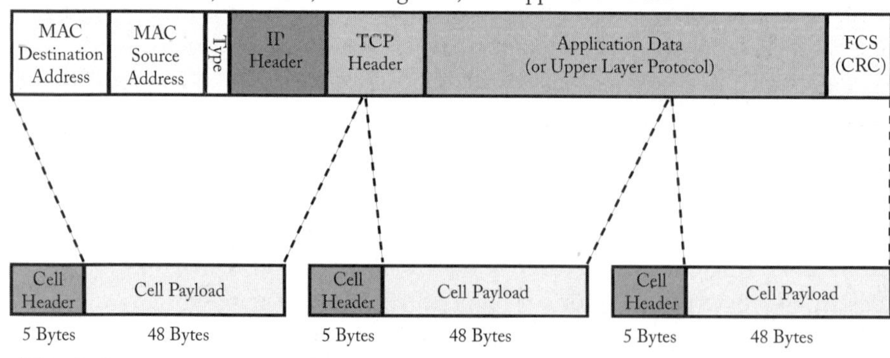

Figure 4.3 *Segment/Packet/Frame into Cells*
All ATM cells are 53 bytes in length and consist of a 5-byte header and a 48-byte payload. To be transported within ATM cells, LAN frames must be divided into pieces of 48 bytes or less.

The ATM standards groups defined two different header formats, referred to as the User-Network Interface (UNI) and Network-Network Interface (NNI). UNI provides communications between ATM endpoints (workstations, hosts, and routers) and ATM switches within a network. NNI provides communications between ATM networks. While both header formats consist of 5 bytes, the UNI cell header defines six fields whereas the NNI header defines five fields.

LOGICAL LINK CONTROL AND SNAP

Standards bodies such as ANSI and IEEE historically have worked together with the ISO to develop flexible standards. The OSI protocol rules required ANSI/IEEE standards to define a separate common protocol to interface with upper layer protocols. 802.2 Logical Link Control (LLC) was defined to satisfy these requirements (see Figure 4.4).

The established upper-layer protocols of the day used DIX ESPEC2 frames. ESPEC2 frames provide a 2-byte Type field that defines the upper-layer protocol contained within the frame. IEEE 802.3 frames replaced the Type field with a Length field and depended on the addition of IEEE 802.2 LLC to define the upper-layer protocol contained within the frame. To perform this task, 802.2 LLC provides a pair of 1-byte destination and source SAP fields. Like the 2-byte Type field, the Logical Link Control SAP (LSAP) fields contain values that define specific upper-layer protocols. However, because the LSAP values are completely different from the Type field values, the new 802.3 standard frame was incompatible with most established protocols (i.e., IP, DECnet, AFP, XNS).

To solve the problem created by three standards bodies, another standards body, the Internet Engineering Task Force (IETF), salvaged the situation with the development of the LLC Sub-Network Access Protocol (SNAP) extension. SNAP replaces the missing Ethernet Type field and adds the OUI (see Figure 4.5). Although SNAP is

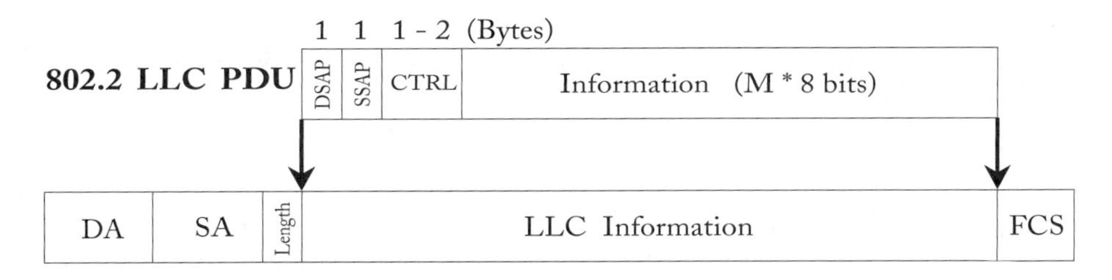

Figure 4.4 *IEEE 802.2 Logical Link Control*
802.2 LLC Protocol Data Units (PDU) may be used with all IEEE 802 LAN frames.
LLC is not used directly with DIX Ethernet frames.

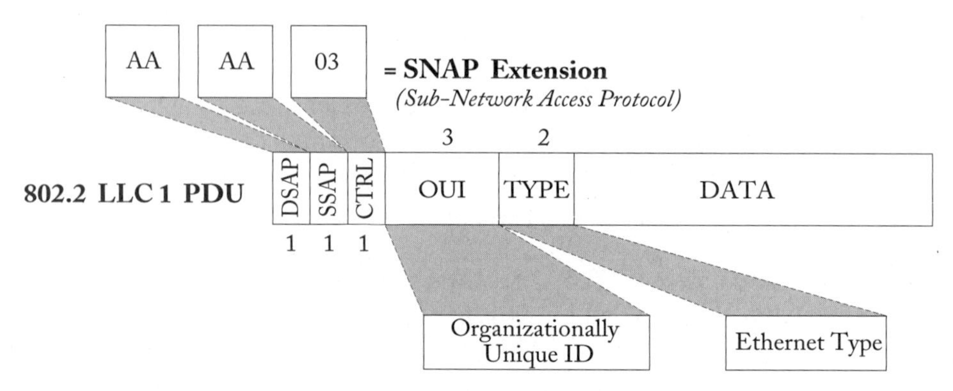

Figure 4.5 *IEEE 802.2 LLC with IETF SNAP Extension*
SNAP replaces the Ethernet Type field omitted by IEEE LAN frames.

rarely used with Ethernet today, 802.2 + SNAP is used to support IP over Token Ring Network and FDDI. AppleTalk requires SNAP for all supported LANs.

NETWORK PROTOCOLS AND APPLICATIONS

As mentioned above, protocols exist throughout all of network communications. LAN access protocols have been discussed in some detail, and other protocols and concepts have been briefly mentioned without much explanation. The following section will elaborate on these all-important protocols and concepts surrounding them. In addition, the role of network applications, operating systems, network operating systems, and their relationships to various protocols will be addressed.

NETWORK COMMUNICATION PROTOCOLS: TCP/IP, IPX, DECNET, APPLETALK ET AL.

One of the key elements to networking is network communication protocols, or simply network protocols, which provide the end-to-end communications and data transport between computers. These protocols correspond to layer 3 of the ISO OSI model, the Network layer. Network protocols come in many varieties to satisfy a range of requirements. In addition to standards-based protocols, just about every computer manufacturer has developed its own suite of protocols at one time or other. There are at least eight different network protocol suites, but only one that is recognized formally as an international standard. The one international standard protocol suite is the ISO OSI, which no one actually uses. And then there is TCP/IP, the one de facto standard protocol suite used by the Internet and which has been adopted worldwide as the multi-platform interoperability solution. Indeed, virtually all computer and networking manufacturers now support TCP/IP, and with it, Internet access, support for electronic mail (e-mail), the World Wide Web (WWW), e-commerce, instant messaging, and all other things Internet-related.

Other networking protocols are still in use but they have all been relegated to serving the requirements of a specific proprietary niche application or platform. Examples of these protocols include Apple Computer Corporation AppleTalk, Banyan Systems VIP, Digital DECnet, IBM SNA, Novell IPX, and Xerox Corporation XNS. Some of these protocols are probably completely extinct by now, having been replaced by TCP/IP or some other protocol (which eventually will also be replaced by TCP/IP). Having been acquired by HP, Digital—once the second-largest computer company in the world, formerly known as Digital Equipment Corporation (DEC)—is no longer in existence. However, a large installed base of DEC computer products continues in service with support from HP/Compaq. DEC developed its own proprietary network protocols that are part of the Digital Network Architecture, or DECnet.

A key feature that differentiates network protocols is the ability to support *routing* across multiple networks. It is important to note that while most network protocols are routable, some are not. Those that are routable support one or more routing update protocols that automatically propagate routing information between network routers. Over the years many network protocols have been developed and even flourished for a while, only to have been culled by economic forces and the momentum of TCP/IP.

Table 4.1 (on the next page) lists network protocols along with their associated routing protocols, the vendor or standards body responsible for the protocol's development, and the protocol's current status. Status for each protocol is identified as follows: current (currently used), defunct (no longer used), EOL (end of life, phasing out), or special (used only for certain applications).

Digital was not known for providing the best TCP/IP support. This created a void that was filled by several third-party companies. The same was true in the early days of Microsoft Windows. Before the release of Windows NT v.3.51 and Windows 95, Microsoft products did not include support for TCP/IP. In the early days of Windows v.3.*x* and Windows for Workgroups v.3.11, accessing the Internet required third-party software in the form of a TCP/IP protocol stack and suite of applications and utilities developed to work with that particular protocol vendor's TCP/IP stack. If some other feature, application, or utility was required, you had to write it yourself or go without. This all changed in 1993 with the release of a standard application program interface (API) for TCP/IP called Windows Sockets (WinSock v.1.1). WinSock provides a simple consistent API to which software developers can easily write new TCP/IP-based applications and utilities. Today all Windows-based TCP/IP stacks include support for the WinSock API, and virtually all TCP/IP-based Internet applications written for Windows take advantage of the WinSock API.

Virtually all contemporary operating systems include native support for TCP/IP. This includes Apple's most current operating systems, all versions of Microsoft Windows, all current versions of Unix and Linux, as well as current IBM systems and Novell NetWare since version 5 (while version 4 did support TCP/IP, it did so very

TABLE 4.1

Protocol	Routing Protocols	Developer/Standards Body	Status
IP	RIP1, RIP2, OSPF	IAB/IETF	current
	(IGRP), E-IGRP	Cisco	current
IPX	RIP, NLSP	Novell (Netware)	EOL
DECnet/IV	DRP	Digital	EOL
DECnet/OSI	IS-IS	Digital	special
DDP	RTMP	Apple (AppleTalk)	EOL
VIP	VRTP	Banyan (Vines)	defunct
CLNP	IS-IS	ISO	defunct
XNS	RIP	Xerox	defunct
NetBIOS	non-routable	IBM, others	EOL
NetBEUI	non-routable	Microsoft (Windows)	EOL
DECnet/LAT	non-routable	Digital	EOL
SNA	non-routable	IBM	current

poorly using encapsulation and it was anything but native to the operating system). Today TCP/IP is also appearing in cell phones, personal digital assistants (PDAs), and other appliances.

TCP/IP APPLICATIONS AND UTILITIES

TCP/IP typically requires specific applications to do anything over a network connection. For example, copying files between computers requires a File Transfer Protocol (FTP) server application on one computer (called a *daemon* in Unix parlance—FTPd), and an FTP client application on the other. To view a Web page requires a Hypertext Transfer Protocol (HTTP) server application on one computer to host the Web page, and a Web browser application on another computer to view the Web page. It is technically possible for any computer that supports TCP/IP to be a server, client, or both for any TCP/IP application. TCP/IP represents an entire suite of network protocols and applications, the details of which fill many volumes. This text will provide only enough information about TCP/IP to explain specific networking issues.

NETWORK OPERATING SYSTEMS

What makes a network operating system (NOS) different from an operating system (OS)? Today the answer is: very little. However, all operating systems were not always network-aware, and just because an OS was network-aware did not necessarily qualify it as a network OS. The first NOS products to emerge included Novell NetWare and Banyan Vines. These operating systems were engineered to provide network-specific functionality and services such as shared access to large disk drives, printers, and multi-user applications, and provided additional security that most mere operating systems lacked. These network operating systems were either not intended to support, or were incapable of supporting, client/user applications. Most were/are dedicated to providing network services, hence the term "dedicated server." By this definition, since Vines is defunct, NetWare is the only remaining NOS.

Some may also insist that some versions of Windows, Unix/Linux, and Apple's Mac OS are in fact network operating systems. While all current releases of these operating systems are certainly network-aware, they are by no means dedicated to supporting network services only. Of these, only Microsoft offers a server-specific release of its Windows operating system(s); however, Windows 2000 Server and Advanced Server share the same operating system kernel as the client workstation version, Windows 2000 Professional. The server versions augment the basic OS with additional networking capabilities and server-related applications and utilities. The new Windows XP Home and XP Professional operating systems are also network-aware. The server version of Windows XP is called Windows 2003. Table 4.2 provides a fairly comprehensive list of operating systems. Table 4.3 lists several common mainframe and mid-range operating systems, and Table 4.4 lists the most common network operating systems.

Access to the services of the NOS is provided by software on the client/user computers. Most such software is currently included in all operating systems. All that is required to use this software is for it to be enabled and properly configured on each computer. This was not always the case with earlier releases of most operating systems. The software consists of a network shell or driver that provides a seamless interface to NOS services as if they were just extensions of the client/user computer. In other words, rather than requiring a special program to copy files to and from the server, the server file system appears as if it were local to each client/user computer. For example, most DOS/Windows computers have their own local A: drive (the floppy), a C: drive (primary hard drive), and perhaps a D: drive (a CD-ROM drive). Client/user computers may access the server file system using additional drive letters E: through Z: (drive wrapping) or via the Universal Naming Convention (UNC). An example of UNC use would be "\\server_name\share_name". In either case, the OS and applications running on the local computer can access the resources of the server NOS as if the resources were installed directly in the local computer.

As operating systems evolved in tandem with networks, the two technologies eventually merged to the point where all contemporary computer operating systems provide some degree of network functionality. Some OS implementations may be limited to mostly client-side or workstation use whereas other implementations of that OS may provide a rich suite of server-side features and applications. Because of this, the arcane term "NOS" is rarely used anymore.

TABLE 4.2 OPERATING SYSTEMS

OS	Vendor	Variants	Status*
AIX	IBM		current
BeOS	Palm, Inc (formerly Be Inc.)		uncertain
Free BSD	Berkeley Unix		current
DOS	Microsoft MS DOS	IBM PC DOS, Novell DR DOS	defunct
HP/UX	Hewlett-Packard		current
IRIX	Silicon Graphics Inc.		current
Linux	public domain	many	current
Mac OS	Apple		EOL
Mac OS-X	Apple		current
Solaris	Sun Microsystems		current
SunOS	Sun Microsystems		defunct
Ultrix	DEC		defunct
Unix	many	many	current
UnixWare	SCO (formerly Novell, formerly AT&T USL)		current
Windows 3.*x*	Microsoft	MS Windows for Workgroups	defunct
Windows 95	Microsoft		EOL
Windows 98	Microsoft		EOL
Windows Me	Microsoft		current
Windows NT	Microsoft	Professional, Server, Adv. Server	EOL
Windows 2000	Microsoft	Professional, Server, Adv. Server, Datacenter Server	current
Windows 2003	Microsoft	Datacenter Edition, Datacenter Edition for 64-bit Itanium 2 Systems, Enterprise Edition, Enterprise Edition for 64-bit Itanium 2 Systems, Standard Edition, Web Edition, Small Business Server 2003	current
Windows XP	Microsoft	Home, Professional	current

*EOL = End Of Life

TABLE 4.3 MAINFRAME AND MIDRANGE OPERATING SYSTEMS

OS	Vendor	Platform	Status
OpenVMS	HP	VAX, µVAX, and Alpha	current
OS/400	IBM	Midrange iSeries (AS/400)	current
OS/390	IBM	Mainframe	current
VM	IBM	Mainframe	defunct
VMS	DEC	(now HP OpenVMS)	current
z/OS	IBM	Mainframe	current
z/VM	IBM	Mainframe	current

TABLE 4.4 NETWORK OPERATING SYSTEMS

OS	Vendor	Variants	Status
3Plus Share	3Com		defunct
3Plus Open	3Com		defunct
LAN Manager	Microsoft	3Com 3Plus Open	defunct
MS Net	Microsoft	IBM PC Net, 3Com 3Plus Share	defunct
NetWare	Novell		current
PC Net	IBM		defunct
Vines	Banyan		defunct

CHAPTER SUMMARY

Binary is the fundamental building block of all data processing and data communications. Bits are assembled into bytes, which are grouped to form meaningful information. From the applications and operating systems, through the network protocols and hardware, and onto the network medium, each step in the data communications process manipulates bits and bytes. Applications pass streams of data to network protocols, which segment the data into manageable pieces and encapsulate it in packets. Those packets are then passed to the network interface where they are encapsulated into frames for transmission onto the network medium. Network operating systems, network protocols, and associated applications facilitate the exchange of information between computers.

REVIEW QUESTIONS

1. Define bits and bytes and the relationship between them.
2. Convert the binary value 11110000 into its decimal equivalent.
3. What is the decimal value of one kilobyte?
4. Define frames, packets, and cells and explain their differences.
5. Identify the data element that corresponds to each OSI layer.
6. Define MAC address and OUI.
7. Describe the relationship between segments, packets, and frames.
8. Define the following acronyms:

OUI	PDU
LLC	SNAP
UNC	WinSock
OS	NOS
FTP	HTTP

9. What makes an OS different from an NOS?
10. Name two network operating systems.

LAN Technologies

OBJECTIVES

After completing this chapter, you should be able to:

- Identify two of the most common local area network access protocols.
- Describe the fields in both 802.3 and DIX Ethernet frames.
- Describe the fields in an 802.5 token ring frame.
- Define CSMA/CD and explain its functions.
- Explain the functions of token passing.
- Describe the fields in an FDDI frame.
- Describe MAC frame addresses.
- Define FDDI and CDDI.
- Define BEB and NVP.
- Define SlotTime.

INTRODUCTION

Learn about the network access protocols used by several of the most popular LAN technologies. Also referred to as network access methods or channel arbitration methods, these protocols govern access to the shared network medium. Each LAN technology uses a unique frame structure to transport data payloads across the network medium. In some cases, elements within the frame structure support the mechanics of the network access protocol. We will also look at the various standards that define each of these LAN technologies and their origins.

ETHERNET

The term Ethernet describes a local area network that supports several different types of media and topologies, and several data rates. All support the same network access method and basic frame structure. The rules for extending a network vary with speed and media.

NETWORK ACCESS METHOD: CSMA/CD

Ethernet network access is governed by the channel arbitration mechanism called carrier sense, multiple access (CSMA). Because multiple stations share access to the common medium, any station attempting to transmit must first listen for the presence of other transmissions (referred to as carrier sense). If no transmissions are under way, carrier is not sensed, and the station may begin transmitting immediately. If a transmission is in progress, carrier should be sensed; the station must defer to that transmission and will continue to check for activity until the network becomes silent. Once the transmission ends and carrier is no longer sensed, a waiting period must elapse before the station may again attempt to transmit. The waiting period is referred to as the *InterFrame Gap* (IFG)and has duration of 9.6μs (for 10 Mbps Ethernet), or about 12 byte times. The IFG for 100Base-T Fast Ethernet is .96μs (see Figure 5.1).

At the beginning of each transmission there is a window of time during which it is possible for two or more stations to begin to transmit at the same time. The result of multiple simultaneous transmissions is called a *collision*. Collision detection (CD) was added to handle collisions after-the-fact and arbitrate access for retransmission, hence the full name of the network access protocol: carrier sense, multiple access, with collision detection, or CSMA/CD. Collisions are normal events in CSMA/CD networks.

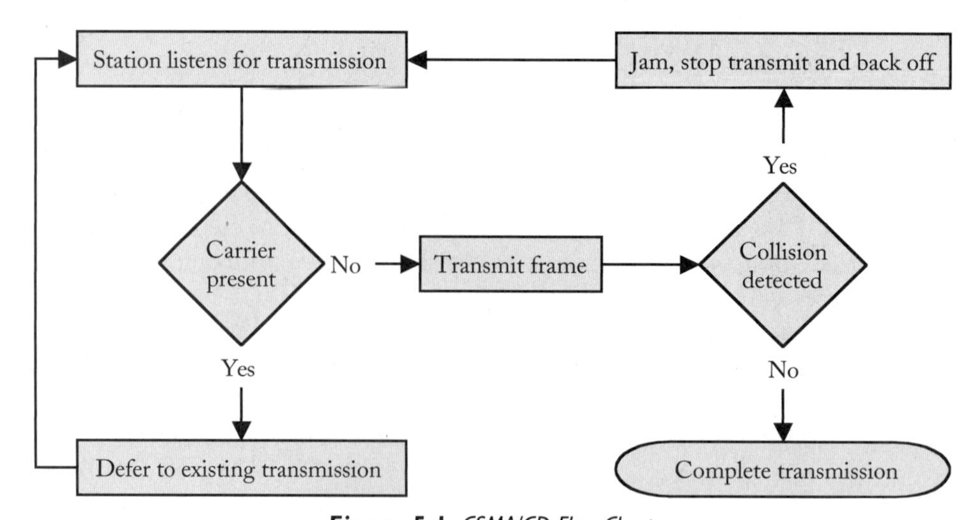

Figure 5.1 *CSMA/CD Flow Chart*

When a collision does occur, the participating stations must stop transmitting data, and transmit a 32-bit *Jam* signal instead (a short stream of "garbage bits" used to enforce the collision event). CSMA/CD then executes the *back-off algorithm*. The back-off algorithm causes the stations participating in the collision to wait a short, pseudo-random length of time before attempting to retransmit. Its purpose is to stop the multiple simultaneous transmissions and act as a traffic cop by first allowing one station to complete its transmission before allowing the other to attempt its transmission. However, unlike the traffic cop example, with CSMA/CD there is no specific device directing traffic flow. The selection of which station gets to transmit first after a collision is detected is determined by the results of the back-off algorithm running on each of the collision participants (see Figure 5.2).

In a properly constructed Ethernet LAN (i.e., one that accurately follows the standards specifications), if a collision is to occur it will happen within the first 512 bits (64 bytes) of the transmission. This number of bits corresponds to an Ethernet time interval called the *SlotTime*, which is based on the time it takes to transmit 512 bits of a frame. The actual time required to transmit 512 bits varies depending on the Ethernet transmission

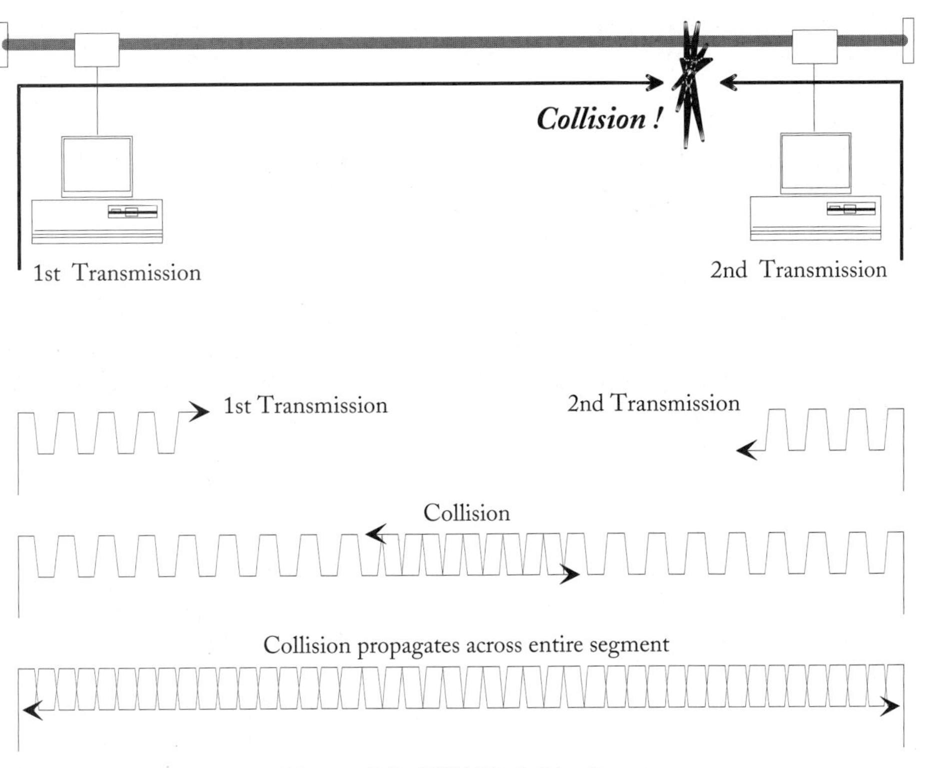

Figure 5.2 *CSMA/CD Collision Event*

speed. For 10 Mbps Ethernet each bit requires 100 nanoseconds to transmit, and for 100 Mbps Fast Ethernet each bit requires 10 nanoseconds. This is known as a *bit time.* And 512 bits translates to 51.2μs for 10 Mbps Ethernet and 5.12μs for 100 Mbps Fast Ethernet. This value corresponds to the round trip propagation delay time—the length of time required for a transmitted frame to reach the farthest possible extent of the Ethernet LAN, encounter a collision, and for that collision event to propagate all the way back to the originating station. The SlotTime also defines the required minimum length of an Ethernet frame as 64 bytes, which equals 512 bits.

These values, combined with the nominal velocity of propagation (NVP) of each type of Ethernet media, define the physical extent of an Ethernet LAN. The NVP describes the speed at which a signal can travel through a given medium and is expressed as a percentage of the speed, or velocity, of light. For example, most Cat5 UTP cabling is rated with an NVP of .595c, or approximately 60% of the speed of light, "c."

Ethernet relies on the truncated Binary Exponential Backoff algorithm (BEB), which sounds far more complicated than it really is. BEB is activated when a collision is detected while a station is transmitting a frame. When a station encounters its first collision for a given frame, BEB generates a random number between 0 and 1, and multiplies the result by 512 (bit times). If the collision is between two stations, and each station generates a different random number, their back-off periods will be different. Of course, for the station that generates a 0, 512x0=0, which means that station may attempt to transmit immediately. For the station that generates a 1, 512x1=512, which means that station must back off for 51.2μs (for 10 Mbps Ethernet). This provides the first station enough time to complete the transmission of at least 512 bits—enough time for the transmission to reach the farthest extent of the Ethernet *collision domain.*

It is also possible that the BEB result after the first collision could result in the two stations generating the same random number. During the first backoff there is an equal probability that BEB will generate a 0 or a 1 (a 50/50 chance). If both stations generate a 1 value, they will both back off for 51.2μs, and if both stations generate a 0 value, they will both attempt to transmit immediately. In either case the result is a subsequent collision. The second collision for a given frame prompts BEB to exponentially increase its random back-off range from 0 or 1, to 0, 1, 2, or 3. By increasing the range of possible multipliers, the probability of another collision is decreased.

If in fact a third subsequent collision does occur for the same frame, BEB will exponentially increase its random back-off value from 0, 1, 2, 3, to 0, 1, 2, 3, 4, 5, 6, 7. Note the *binary exponential* progression from one bit (0 or 1), to two bits (00, 01, 10, 11), to three bits (000, 001, 010, 011, 100, 101, 110, 111), and so on. After the maximum exponential back-off range of 10 bits is reached (a maximum multiplier value of 1024), it is truncated. An error is generated after 16 failed attempts to transmit a given frame (i.e., 16 consecutive collisions in the attempt to transmit a single frame).

CSMA/CD networks are considered to be probabilistic or non-deterministic. The CSMA/CD network access protocol is a random-access contention protocol. In other words, the sender of data does not know if the network is available for its own use until it contends for its use and is successful.

FRAME STRUCTURE

The basic frame structure of both IEEE 802.3 and DIX ESPEC2 Ethernet is the same. The subtle differences between the two specifications permits their coexistence on a common network, but prevents direct interoperability. The minimum size of an Ethernet frame is 64 bytes, which corresponds to the SlotTime used by collision detection. The maximum size of an Ethernet frame is 1518 bytes (see Figure 5.3).

The preamble consists of seven octets of alternating ones and zeros. It provides the clocking necessary for the receiving station to synchronize to the incoming bit-stream.

10101010 10101010 10101010 10101010 10101010 10101010 10101010

The start of frame delimiter (SFD) consists of one octet of 10101011. It indicates the beginning of the MAC frame. In DIX ESPEC2, a discrete SFD field does not follow the preamble; instead, the SFD bit sequence is included in the eight-octet DIX preamble. In other words, the DIX ESPEC2 preamble is the same bit sequence as the IEEE 802.3 preamble plus the SFD.

The destination address and the source address may each be 2 or 6 bytes. Locally administered addressing requires only 2 bytes. Now considered obsolete, it is rarely used and the IEEE may eventually drop it from the specification in favor of universal

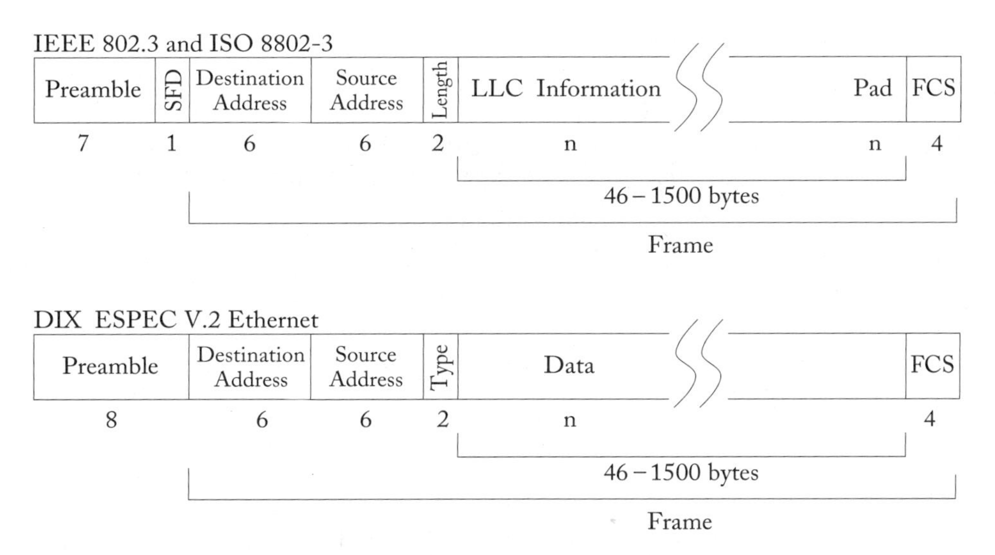

Figure 5.3 *IEEE 802.3 and DIX ESPEC2 Ethernet Frames*

addressing, which uses 6 bytes. The first 3 bytes are called the Block ID, or Organizationally Unique Identifier (OUI), are unique to each manufacturer, and are controlled by the IEEE Registration Authority. The second set of 3 bytes is called the Device ID, is assigned by the manufacturer, and is unique to each piece of hardware (e.g., NIC, switch, and router interface).

IEEE 802.3 specifies a 2-byte length field where DIX ESPEC2 specifies a 2-byte Type field. This is the first significant difference between the two specifications. The DIX Type field value defines the higher layer network communication protocol being used within the data field portion of the frame. The IEEE now governs the values placed in this field (they were once controlled by Xerox Corp.). Almost all Type values begin above 0600h (0x0600). These numbers are expressed in hexadecimal, which will be addressed later in this text.

The IEEE moved the function of the Type field into the Logical Link Control (802.2 LLC) information header so that the value could be passed through the LLC sublayer to the network layer. The IEEE 802.3 uses the length field to define how many octets are actually in the LLC Information field (less any padding bytes applied via the pad field). The largest value permitted by the standards is 05DCh (0x05DC), which represents 1500 bytes in decimal. This corresponds to the maximum length of the entire LLC protocol data unit (PDU), which corresponds to the conventional maximum length of the 802.3/Ethernet data field. Some Ethernet implementations support *Jumbo Frames*, which allows frames up to 9k bytes in length; however, because of the potential for innumerable network interoperability problems, this atypical implementation is not highly recommended.

The LLC Information and ESPEC2 Data fields both carry upper layer protocol information and data. DIX ESPEC2 does not define a discrete LLC sublayer in its specifications, nor does it also does not define a separate pad field. The IEEE 802.3 pad field is used to fill the frame's data field to its minimum of 46 bytes. If the amount of actual LLC data placed in the frame is less than 46 bytes, the pad field will be expanded to compensate. DIX ESPEC2 Ethernet requires upper layer protocols to assume this responsibility.

The frame check sequence (FCS) field size and algorithm are the same for both IEEE 802.3 and DIX ESPEC2 Ethernet. The algorithm is a cyclical redundancy check (CRC) of the contents of the frame that produces a 32-bit result (4 bytes). This value is placed in the FCS field. The receiving device executes the same algorithm and compares its results to the FCS value contained in the frame. If the values match, the frame is accepted; if they do not match, the frame is discarded.

TOKEN RING

The term token ring describes a local area network that supports several different types of media, one basic topology, and several data rates. As with Ethernet, all support the same network access method and basic frame structure. Network extension rules vary with speed and media.

NETWORK ACCESS METHOD: TOKEN PASSING

Token ring network access is governed by the mechanism called token passing. A specialized packet called a *free token* constantly circulates the ring. Stations wishing to transmit on the network must wait for a free token (see Figure 5.4).

The station must then convert the token into a data frame, insert its data, addressing information, and other fields, and then transmit. A data frame generated by a station circulates around the ring, passing through all intermediate stations, to reach its destination station. The destination station must recognize the destination MAC address within the frame as its own. The destination station will then copy the contents of the frame into memory and retransmit the frame onto the network, allowing the frame to return to its source station. As the frame is transmitted back onto the ring, the station modifies the last field to indicate that it recognized the destination address and the frame contents were successfully (or not) copied by the destination station (see Figure 5.5).

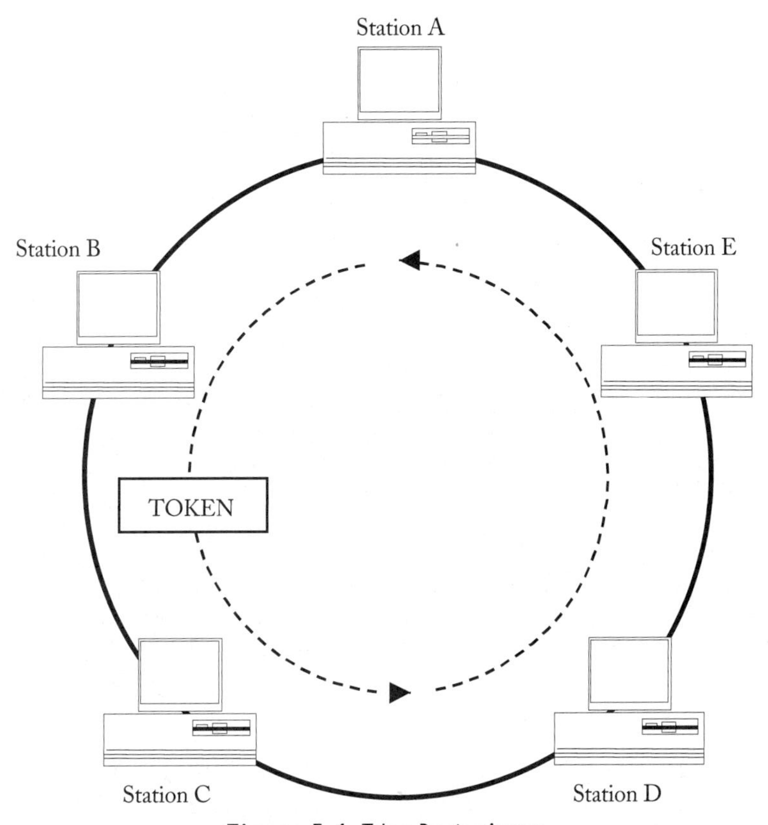

Figure 5.4 *Token Passing Access*

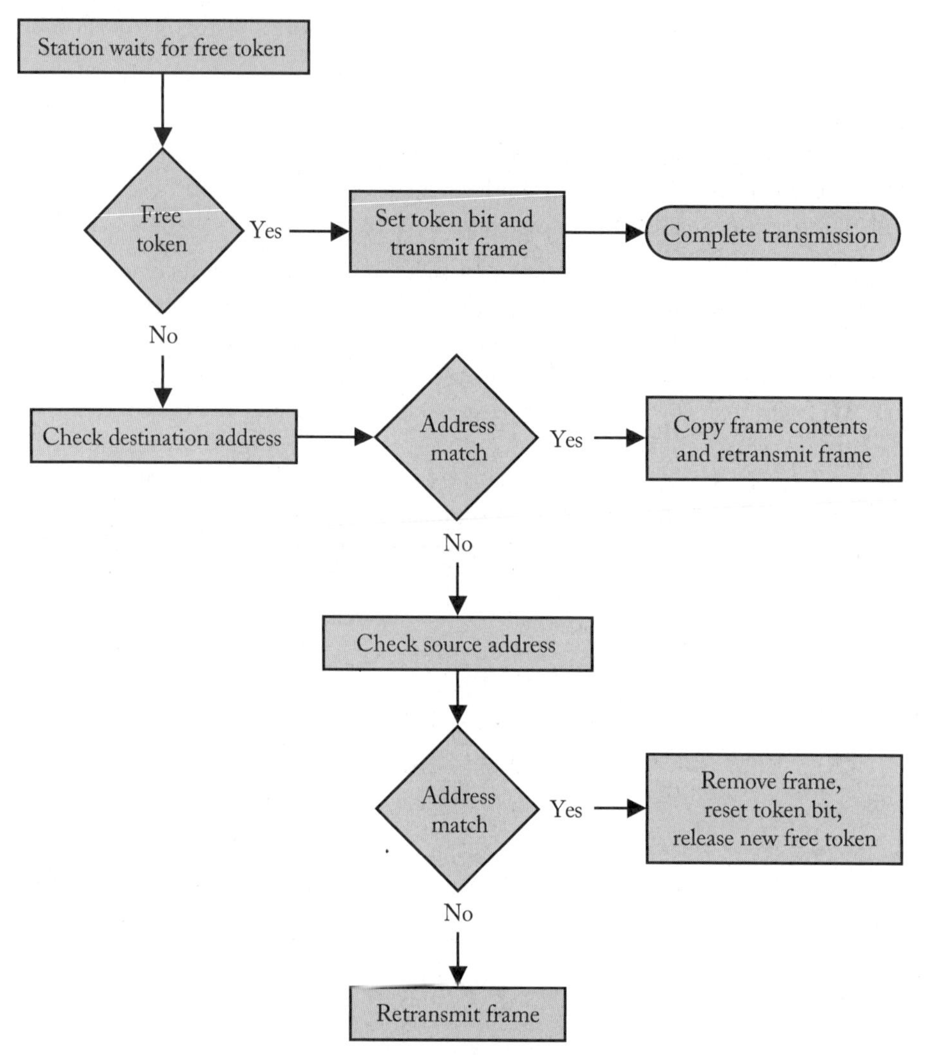

Figure 5.5 *Token Passing Flow Chart*

Once the frame is retransmitted back onto the ring it must continue circulating through all remaining stations between the recipient and the source station. Based on the original 802.5 specifications, when the source station receives the frame it transmitted, it must completely purge the frame from the ring and then release a new free token.

With the release of the 16 Mbps token ring standard, support for early token release (ETR) was added. ETR is an option that increases available network capacity by improving data transmission efficiency. Rather than delaying the release of each new free token until after a frame has been purged from the ring, ETR allows a transmitting station to release a free token as soon as it completes frame transmission. However, use of ETR may disrupt priority reservations and increase the access delay for priority traffic when the network is heavily loaded with small frames. Stations that implement ERT are interoperable with stations that do not.

Token ring supports a Quality of Service (QoS) mechanism called priority operation. In the access control field of all frames and free tokens there are three priority bits and three reservation bits. The 3-bit priority scheme supports eight different priority levels, 0 through 7. Normal operation uses level 0.

Using Figure 5.4 as an example, a station with high-priority data (a) places a reservation request by setting the reservation bits in any passing frame. The source of that frame (b) reads the Reservation request as it purges the frame from the ring in preparation for releasing a new free token. The new token will be released at the requested priority level. At this point, any station on the ring with high-priority data greater than or equal to the priority level of the token may capture that high-priority token and transmit a high-priority frame. Assuming that no other stations are awaiting a high-priority token, the station (a) that requested the priority request will receive the high-priority token. That station will then transmit its high-priority frame, which will circulate around the ring to reach its destination (d). The destination station will copy the contents of the frame, set the frame status bits, and retransmit the frame back onto the ring. The frame will complete its journey around the rest of the ring, passing through all remaining stations until it reaches the source station (a). The source station purges the frame and releases a new free token *at the same high priority*. If no other stations require that high-priority token, it will reach the original station (b) that originally generated the high-priority token. That station will then release a new free token set to normal priority (0).

The intent of priority operation was to provide QoS support for real-time interactive applications such as voice and video conferencing. In this respect, token ring was well ahead of its time (1985); however, operating systems and applications must be capable of supporting the priority operation, and most do not. Recent releases of several popular operating systems include such support, but that support was too late for token ring.

The maximum size of a token ring frame is dependent on the token holding timer (THT). In essence, the THT defines the length of time a given station may transmit. Factors that affect this value include the speed of the ring (4 or 16 Mbps), the overall length of the ring (including all lobe cables, the number of MSAUs and MSAU trunk cables), and the buffer in each token ring network interface card.

The acknowledged frame must circulate through all remaining stations on the ring to reach the source station. The station that generated the data frame must receive and remove it from the ring, and release a new free token in order for another station to use the network. Management mechanisms have been incorporated into the token passing access protocol to monitor the network for errors, maintain normal operations, and handle faults in most parts of the system. The token ring management mechanisms include Active Monitor Present, Lost Token, Circulating Frame, Standby Monitor Present, Neighbor Notification, Token Claiming, and Beaconing. The token ring network interface cards perform these low-level functions without additional support or human intervention.

Each ring requires an Active Monitor. The Active Monitor is a station on the ring that provides most of the ring management functions. Any station on the ring can become the Active Monitor, but there can be only one Active Monitor at any given time.

Tasks performed by the Active Monitor:

- Initiates Active Monitor Present and Neighbor Notification procedures.
- Ensures that only one frame or token is circulating on the ring at any time.
- Ensures that there is always either a frame or a token on the ring.
- Prevents frames from circulating around the ring more than once.
- Checks for the presence of other Active Monitors.

Active Monitor Present: Approximately every 7 seconds the Active Monitor station transmits an *Active Monitor Present (AMP)* frame. The AMP frame circulates around the ring to notify all other stations of the identity and presence of the Active Monitor.

Lost Token: Each time the Active Monitor sees the starting delimiter of a frame or token, it resets a timer. If the timer expires before the Active Monitor sees another frame or token, it issues a *purge frame* to clear the ring. The Active Monitor then issues a new free token.

Circulating Frame: When standby monitor station transmits a frame it sets the *monitor bit* in the *access control* field to 0 (M=0). When the Active Monitor station transmits a frame, it sets the monitor bit in the access control field to 1 (M=1). If the Active Monitor sees a frame other than its own with the monitor bit set to 1, it issues a purge frame and issues a new free token.

Tasks performed by all Standby Monitors:

- Monitor the status of the Active Monitor.
- Participate in Standby Monitor Present transmissions.
- Participate in Neighbor Notification.
- Ensure that the token passing mechanism is functioning.
- Assume the role of Active Monitor should the need arise.
- Initiate Beaconing in the event of loss of network signal.

As mentioned above, each ring requires one Active Monitor; all other stations on the ring are *Standby Monitors*. Standby Monitors perform a set of tasks including monitoring the status of the Active Monitor. If the current Active Monitor fails for some reason (e.g., the network cable is disconnected or the station is powered down), the Standby Monitor with the highest MAC address will assume the role of Active Monitor.

Standby Monitor Present: When each Standby Monitor sees an AMP frame, it prepares to transmit a Standby Monitor Present (SMP) frame. Each Standby Monitor queues its SMP frame as it awaits a free token. Each Standby Monitor must wait for a free token to transmit its SMP frame. This results in a succession of SMP frames being transmitted after every AMP frame (about every 7 seconds).

Neighbor Notification: Each station on the ring maintains a list of the MAC addresses of the other stations on the ring. The nearest upstream neighbor is flagged after every AMP/SMP transmission. This allows a station to know the MAC address of the station to which it listens, but it cannot know the MAC address of the station to which it transmits.

Token Claiming: Each Standby Monitor resets a timer after receiving an AMP frame. If the timer expires before the Standby Monitor sees another AMP frame, it issues a *claim token* frame. Again, the Standby Monitor with the highest MAC address will assume the role of Active Monitor, issue a purge frame to clear the ring, and issue a new free token.

Beaconing: When any station including the Active Monitor detects a loss of signal (defined as no signal for 7 seconds), it transmits a *beacon frame* around the ring to its upstream neighbor. Each station learns the MAC address of its upstream neighbor through neighbor notification (discussed above). Since stations can only transmit in one direction around the ring, the only way a station can communicate with its nearest upstream neighbor is by transmitting a frame all the way around the ring. When the upstream neighbor receives the beacon frame, it removes itself from the ring and performs a lobe media test to determine if there is a fault somewhere in its connection to the MSAU. If that station passes its lobe media test, it rejoins the ring and Beaconing continues until the problem is isolated, the ring stabilizes, or a human operator intervenes.

FRAME STRUCTURE

The basic frame structure of both IEEE 802.5 and IBM Token Ring is the same. The subtle differences between the two specifications may permit interoperability, but does create compatibility problems in token ring environments that use source route bridging. The maximum size of a token ring frame varies depending on speed. For 4 Mbps token ring, the maximum frame size is (approximately) 4092 bytes, while 16 and 100 Mbps token ring support a maximum frame size as large as 17,800 bytes. Maximum frame size is actually dependent on the token holding timer (THT) (see Figure 5.6).

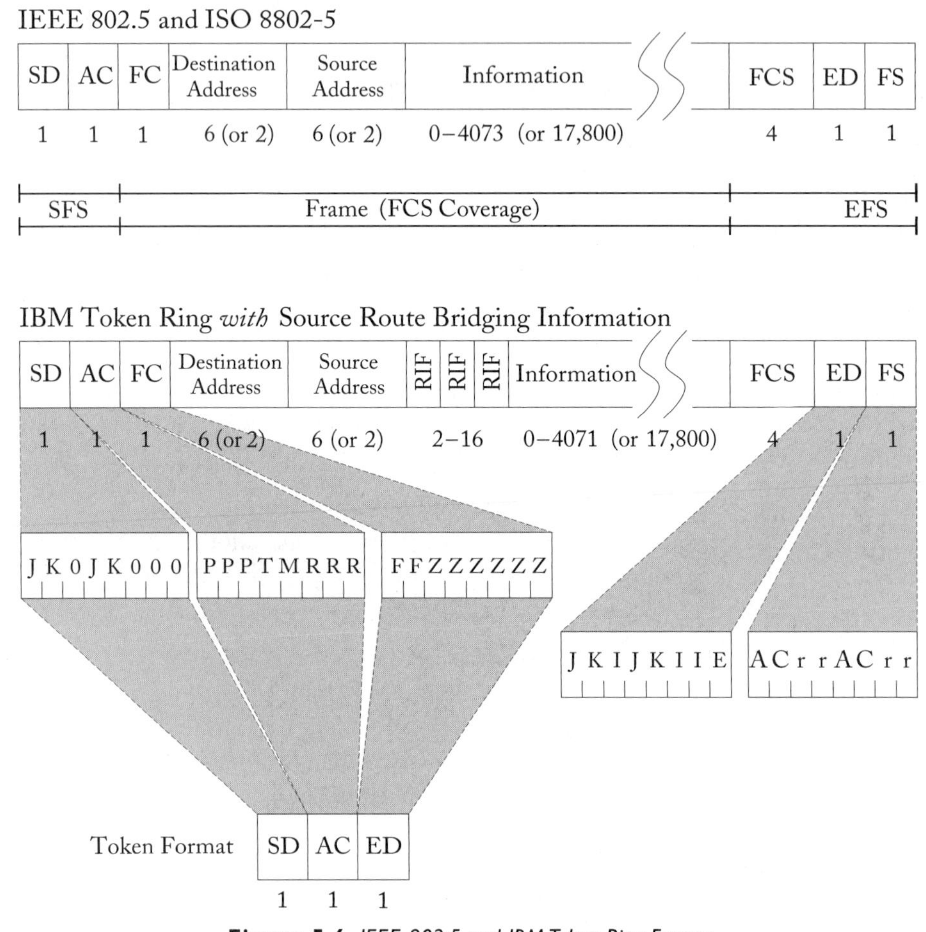

Figure 5.6 *IEEE 802.5 and IBM Token Ring Frames*

As the name implies, token ring relies on specialized frames called *free tokens*. The token format consists of three single-octet fields: the starting delimiter (SD, 1 octet), access control (AC, 1 octet), and ending delimiter (ED, 1 octet). This 3-byte token is the means by which the right to transmit is passed from one station to the next.

The IEEE 802.5 frame is divided into three sections consisting of a total of nine fields. The three sections are the start of frame sequence (SFS), the portion of the frame included in the frame check sequence (FCS) calculation (CRC-32), and end of frame sequence (EFS).

The start of frame sequence consists of two single-octet fields: the starting delimiter (SD) and access control (AC). The starting delimiter contains special non-data "J" and

"K" bits. These bits do not represent characters of the alphabet; the non-data "J" and "K" bits are actually violations of Manchester encoding and are used only to provide timing. The access control field consists of 3 priority bits, 1 token bit, 1 monitor bit, and 3 reservation bits. The priority bits are used to identify the priority level of the current token or frame. The reservation bits are used to request a token at a specific priority level. The token bit identifies whether the current frame is actually a free token or is currently in use. The monitor bit is used by the Active/Standby Monitor present function.

The portion of the frame covered by the frame check sequence consists of four fields: frame control (FC, 1 octet), destination address (DA, 2 or 6 octets), source address (SA, 2 or 6 octets), and the information field (INFO, zero or more octets), which carries upper layer protocol information and data. The frame control field contains an array of network maintenance values and vector information.

The end of frame sequence consists of three fields: frame check sequence (FCS), which contains the results of the CRC algorithm, the ending delimiter (ED), and frame status (FS), which provides a low-level acknowledgement. (Some standards documents indicate that the FCS field is not actually part of the EFS, but rather included in that portion of the frame covered by the FCS. Since the result of the CRC-32 cannot be included in the CRC calculation, this minor point seems contradictory; therefore, this writer has included the FCS field in the EFS.) As with the starting delimiter, the ending delimiter contains special non-data "J" and "K" bits. The E represents the error bit, which may be set anywhere along the frame's journey around the ring. If a frame is returned to the source station and the E bit is set, that frame must be retransmitted. The frame status field provides a low-level acknowledgment that the frame successfully reached its destination. This field contains a pair of address bits (A) and a pair of copy bits (C); the "r" bits are merely reserved (for what, who knows?). When the destination station receives a frame, it must set both address bits to acknowledge that it has recognized the frame's destination MAC address as its own. It must also set both of the copy bits to acknowledge that it has successfully copied the contents of the frame, and then retransmit the frame onto the ring so that it can circulate through all remaining stations and return to the source station. If a frame is returned to the source station and the A and C bits are not set properly, that frame must be retransmitted.

A special bridging protocol used exclusively in token ring networks is called *source route bridging (SRB)*. SRB adds from 2 to 16 octets of route information field (RIF) data. This information is used only by source route bridges to determine the best path for one station to communicate with another across token ring LANs interconnected by source route bridges. Virtually all non-IBM applications do not use SRB and have no means to process the RIF information. SRB was always very rare and is now obsolete.

FDDI AND CDDI

FDDI and CDDI describe a local area network that supports several different types of media, a somewhat complex topology, and a single data rate. Both support the same network access method and frame structure. Network extension rules vary with media and implementation.

NETWORK ACCESS METHOD: TOKEN PASSING

FDDI defines a 100 Mbps fault-tolerant dual counter-rotating fiber optic token passing ring. As the name implies, FDDI uses optical fiber media. It is fault-tolerant because it supports dual fiber rings over which data is transmitted in opposite directions (referred to as counter-rotating). The dual ring topology consists of a primary ring and a secondary ring. During normal operation only the primary ring is used for data transmission and the secondary ring remains idle. The objective of the dual ring architecture is to provide greater reliability and robustness. Should the ring fail in one location (e.g., a broken fiber) the FDDI switches, or nodes, on both sides of the fault will bypass the problem by connecting one loop of the ring onto the other loop (see Figure 5.7).

The Copper Data Distributed Interface (CDDI) began as one vendor's solution to providing FDDI over twisted-pair copper. The first implementations supported STP. Support for UTP was later added, but the only UTP available at the time was voice-grade DIW24, which limited the cable length to about 50 meters. When higher performance UTP became available, it was used to extend the CDDI cable length to 100 meters. CDDI eventually became the industry-standard term for the copper implementation of FDDI.

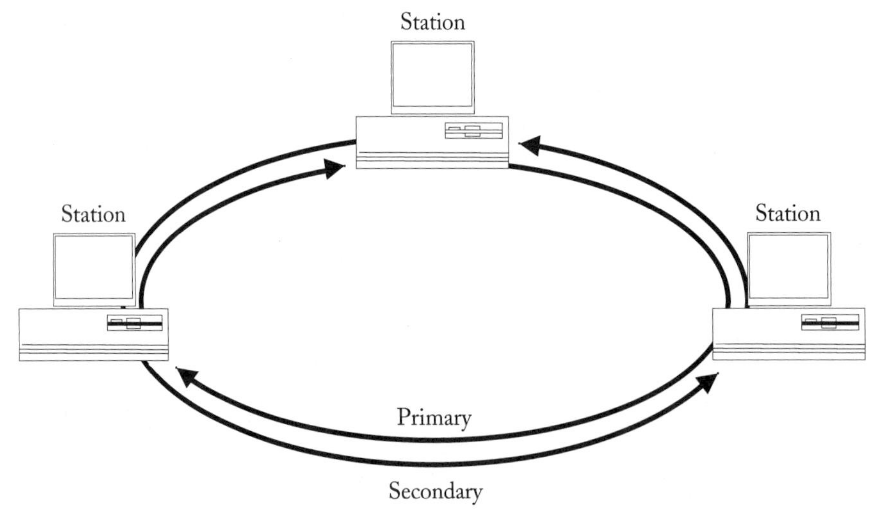

Figure 5.7 *FDDI Counter Rotating Rings*

The Fiber Distributed Data Interface (FDDI/CDDI) uses a token passing network access protocol very similar to IEEE 802.5 token ring. FDDI/CDDI also supports the same 6-byte MAC addressing convention as that used in Ethernet and token ring. Each FDDI frame consists of 32 bytes of framing information, supports a maximum frame size of 4500 bytes, and specifies a minimum frame size of 128 bytes. FDDI/CDDI tokens are special 3-byte frames with a preamble.

FRAME STRUCTURE

The ANSI X3T12 FDDI and CDDI frame format is similar to the format of an IEEE 802.5 token ring frame. The maximum size of an FDDI frame is 4,500 bytes. Figure 5.8 shows the frame format of an FDDI data frame and token.

The preamble (PA) consists of a minimum of 2 bytes, which contain 16 "symbols" of *Idle*. It provides a unique sequence that prepares each station to receive the frame that follows. Subsequent repeating devices throughout the ring may change the length of the Idle pattern consistent with FDDI physical layer clocking requirements. Therefore, repeating devices may generate a preamble that varies in length, either shorter or longer, than the originally transmitted preamble. If for any reason an FDDI device cannot correctly repeat a frame, then it will not repeat any part of the frame (including the starting delimiter).

As with token ring, the starting delimiter (SD, 1 byte) contains special non-data "J" and "K" bits; however, in FDDI these symbols will be seen only in the start delimiter of a frame or token. As with other LAN technologies, this field indicates the beginning of a frame by employing a signaling pattern that differentiates it from the rest of the frame.

Frame control (FC, 1 byte) indicates the size of the address fields, defines whether the frame contains asynchronous or synchronous data, and provides other control information.

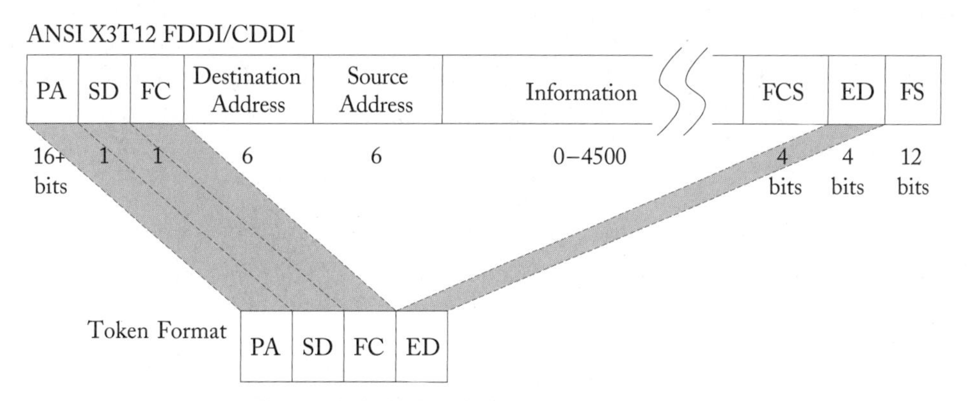

Figure 5.8 *ANSI X3T12 FDDI and CDDI Frames*

The next three fields include the destination address (DA, 6 octets), the source address (SA, 6 octets), and the information field (INFO, zero or more octets), which carries upper layer protocol information and data.

Frame check sequence (FCS) is filled by the source station with a calculated cyclic redundancy check value dependent on frame contents (as with token ring and Ethernet). The destination address recalculates the value to determine whether the frame was damaged in transit. If so, the frame is discarded.

The end delimiter (ED, one 4-bit symbol) consists of a single "T" symbol. The T symbol indicates that the frame is complete. Any frame sequence that does not end with this T symbol is not considered a valid frame and is discarded.

As with token ring, the frame status field (FS, three 4-bit symbols) provides a low-level acknowledgement that allows the source station to determine whether an error occurred in transmission by acknowledging that the frame was recognized and copied by a receiving station.

ALTERNATIVE ETHERNET NETWORK ACCESS PROTOCOLS

To alleviate congestion problems in Ethernet, to address potential problems such as *Channel capture*, and to support multimedia LAN traffic, several new protocols have been proposed especially for Ethernet. Channel capture is an issue that may or may not prove to be problematic. It has been identified as a phenomenon that can generate lengthy access delays and cause the backoff algorithm in CSMA/CD to handle contention in a less than fair manner under certain conditions. An alternative backoff algorithm called BLAM (Binary Logarithmic Arbitration Method) has been proposed as a solution. In most cases, as you will see, Ethernet switches are required to provide the best solutions.

802.3w–BLAM

The Binary Logarithmic Access Method is intended as a potential alternative to Ethernet's existing truncated Binary Exponential Backoff algorithm (BEB). Rather than replacing the current system, BLAM would coexist with BEB to provide backward compatibility to legacy LAN systems. Observations of certain adverse effects of Ethernet behavior on networks using high-performance workstations and servers were the driving force behind the development of this alternative algorithm. The use of Ethernet LAN switches can negate these adverse effects.

802.3x–FULL DUPLEX/FLOW CONTROL

Flow control allows vendors to build switches with finite, limited memory, and to prevent packet loss by issuing flow control commands to the attached stations. Pushing the problem back to the station is much preferable to throwing away the packet in the switch due to buffer overflow.

The problem is especially bad when supporting full duplex transmissions, as there is no way to provide "back-pressure" to the station by asserting false carrier or false collisions, since the station ignores these when operating in full duplex. If, for example, you have a lot of workstation ports all sending traffic to a single-server port, there is nothing the switch can do to prevent becoming congested. The idea of flow control is to stop the senders from sending more traffic when this condition occurs rather than just discarding frames.

Flow control does not replace CSMA/CD, rather it is in addition to it. The intent of 802.3x is to implement flow control using standard Ethernet frames with a special Type field value. There is no change to the underlying access protocol and no impact on the low-level hardware. The older ESPEC2 Type field format was chosen to avoid having switches implement 802.2 Logical Link Control layer functions. To accommodate 802.3x, the IEEE has adopted a proposal allowing the older DIX ESPEC2 Type fields in a general manner for use in 802.3 LANs. 802.3x flow-control information is transmitted using the Type field in DIX ESPEC Ethernet frames.

FDDQ–FAIR DUAL DISTRIBUTED QUEUING

FDDQ would supposedly provide utilization and average latency comparable with those of CSMA/CD, but puts packets into one of two global distributed queues when the network becomes congested. In this way, FDDQ would provide two-priority access levels to the network.

PACE–PRIORITY ACCESS CONTROL ENABLED

Developed by 3Com Corporation, PACE is not part of any IEEE working group. It is a proprietary switch-based technology that makes use of installed PCs, workstations, Ethernet adapters, cabling, and management tools, and is intended to preserve investments in administrative and management expertise. PACE technology addresses the problem of variable network delay through a variety of techniques to regulate Ethernet timing and deliver high-quality, real-time multimedia.

An Ethernet switch with PACE employs traffic control algorithms that purportedly allow each link to the switch to operate at more than 98% efficiency, even under full load and when servicing a mix of real-time and conventional data traffic. The traffic control algorithms also provide predictable LAN transmission on Ethernet. The same enhancements work at both 10 Mbps and the Fast Ethernet speed of 100 Mbps.

CHAPTER SUMMARY

Ethernet, token ring, and FDDI are the three dominant LAN technologies of the last fifteen years. While FDDI held promise, the requirements of its dual fiber ring topology were simply too expensive for most organizations to seriously consider. When first released, FDDI and CDDI were the only 100 Mbps LAN standards available, making them reasonable options at the time for situations that demanded LAN data rates in excess of what Ethernet and token ring had to offer.

All LAN technologies use variable-length frames and they all support different minimum and maximum frame sizes. Ethernet supports a relatively small frame size of 1518 bytes, whereas FDDI and token ring support much larger maximum frame sizes of 4500 and 4092 or 17,800 bytes, respectively. But just because a network can support extremely large frames does not mean the network protocols or applications will use them. And more importantly, just because a network can support large frames does not mean it is a good idea. The use of big variable frames causes big variable latency, better known as *jitter*, which can severely affect the performance of interactive applications such as video teleconferencing and Voice over IP.

Debates have raged over the inefficiencies of one network access protocol and the benefits of another. Such protocols are really only meaningful in shared media environments (i.e., LANs that use repeaters/hubs or MSAUs). Now that LAN switches have become so economical, there are no compelling reasons to build a shared media LAN. As switches replace hubs, the shared media concept disappears because each station is now provided with its own dedicated connection. Since each switch connection is dedicated to one station, it obviates the need for channel arbitration. And since the media is no longer shared, full duplex operation may be enabled, further improving throughput. The advent of LAN switches silenced most of the old arguments over network access protocols—the war between Ethernet and token ring has been fought, and switching won.

REVIEW QUESTIONS

1. What is the network access protocol used by Ethernet?
2. Diagram the fields in an ESPEC2 Ethernet frame.
3. Diagram the fields in an IEEE 802.3 frame.
4. Diagram the fields in an IEEE 802.5 frame.
5. What are the minimum and maximum Ethernet frame length?

6. Describe the functions of CSMA/CD.
7. Describe MAC frame addresses.
8. Define the following acronyms:

DIX	IEEE
IFG	CSMA/CD
BEB	NVP
SFD	FCS
AMP	SMP

9. Define SlotTime.
10. Define BitTime.

Classic Ethernet

OBJECTIVES

After completing this chapter, you should be able to:

- Explain the difference between DIX ESPEC2 and IEEE 802.3.
- Define 10Base5, 10Base2, 10Base-T, and 10Base-F.
- Explain the Ethernet transceiver functions.
- Define the purpose of SQE, Heartbeat, and Jam.
- Explain the difference between MDI and MDI-X.
- Define Link Segment, Link Pulse, Link Status, and Null Link.
- Describe the problem known as split pairs and possible causes.
- Explain the difference between 568-A and 568-B connector pin-outs.
- Describe four types of LAN media and the standards that use them.
- Explain the meaning of structured cabling system.

INTRODUCTION

Learn about the media used by the original implementation of the most popular family of LAN technologies: Ethernet. Referred to as Classic Ethernet, Legacy Ethernet, or just plain Ethernet, this chapter will cover the 10 Mbps LAN standards initially supported by DIX ESPEC2 Ethernet and later standardized as IEEE 802.3. Now published by the ISO, the full title of the standards document is International Standard ISO/IEC 8802-3 ANSI/IEEE Standard 802.3 CSMA/CD LANs. The word "Ethernet" defines the original DIX implementation and does not appear anywhere in the standards documents, but that has not stopped anyone from referring to the technology in all its forms as Ethernet!

10 MBPS ETHERNET OVERVIEW

Ethernet is based on an early radio-based network called ALOHAnet. As the name implies, it originated in Hawaii. ALOHA was first implemented in 1969 as a wireless terrestrial network linking several islands using taxicab radios; later, in 1971–72, it was deployed using satellites.

The original specifications for a copper cabling–based 10 Mbps Ethernet were conceived at Xerox PARC and then further developed by Digital, Intel, and Xerox (DIX). Pronounced "ee-thur-net," Ethernet is a trademark of Xerox Corporation USA. The only media supported by this early specification was thick coax or "ThickNet," which later became known as IEEE 802.3 10Base5.

Term such as *10Base5* actually define three pieces of information: the data rate in megabits per second (10), the transmission type (Baseband), and either the type of media or the segment length in hundreds of meters (500). Although the Ethernet frame structure underwent a few changes on its way to becoming an IEEE standard, the media specifications for thick coax remained the same. The specifications support a maximum of 1024 stations in a single Ethernet LAN, or Collision Domain. The IEEE later added support for new and different types of media:

1Base5	1 Mbps, baseband, over 500 meters of UTP phone wire (also known as StarLAN)
10Base5	10 Mbps, baseband, over 500 meters of thick coax
10Base2	10 Mbps, baseband, over 185 meters of thin coax (almost 200 meters)
10Broad36	10 Mbps, broadband, over 3600 meters of CATV coax (defunct)
10Base-T	10 Mbps, baseband, over ~100 meters using 2 pairs of DIW24 UTP
FOIRL	10 Mbps, baseband, over 1 km using two strands of optical fiber
10Base-FL	10 Mbps, baseband, over 2 km using two strands of multimode optical fiber

You may have noticed that the IEEE has not been a stickler for consistency when it comes to naming standards. For some reason the fiber optic inter-repeater link (FOIRL) standard did not adhere to the three-part naming convention; however, the FOIRL has long since been superceded by the 10Base-FL standard. And, of course, the IEEE later developed standards for Fast Ethernet and Gigabit Ethernet including 100Base-TX and 1000Base-T. There was also a 1 Mbps Ethernet implementation called 1Base5, more commonly referred to as *StarLAN*, but it was quickly eclipsed by 10Base-T.

The physical interface to the Ethernet media is provided by a component commonly referred to as a transceiver, defined as a Medium Attachment Unit (MAU) by the IEEE standards. The 802.3/Ethernet MAU has absolutely nothing whatsoever in common with the IEEE 802.5 token ring Multi-Station Access Unit, also called a MAU (or MSAU). Because this component is the common denominator for all Ethernet connections, we will begin with a description of transceiver functions.

TRANSCEIVERS, CABLES, AND CONTROLLERS

All implementations of Ethernet rely on consistent specifications for transceivers, cabling, and network controllers or network interface cards. Transceivers are where the rubber meets the road—or in this case, where the data stream meets the media.

TRANSCEIVER FEATURES AND FUNCTIONS

All Ethernet connections require a transceiver, regardless of media type. Some transceivers are external devices, connected to their host computer or internetworking component via an Attachment Unit Interface (AUI) or drop cable. Others are built into the computer's network interface card (NIC, sometimes referred to as a network adapter or controller), or embedded directly in an internetworking component. Different transceivers are required to provide support for each of the different media (see Figure 6.1).

Ethernet transceivers perform several primary and secondary functions and many vendors have added an array of LED indicators to their transceivers to indicate the status of some of these functions. The following table identifies the three primary and three secondary functions of IEEE 802.3/Ethernet transceivers (see Table 6.1).

TRANSMIT AND RECEIVE DATA

All transceivers are responsible for transmitting data onto and receiving data from the network media. They must also monitor the media for activity and collisions while they are transmitting. All transceivers must be connected to a computer or other data terminal equipment (DTE) device to function. Transceivers cannot be connected directly to each other via the AUI drop cable. Power is supplied to the transceiver from its host device (DTE) via the AUI cable (or circuit). The transceiver's power requirement is +12 volts DC at a maximum of 0.5 amps. Additional information may be found in the ISO/IEC 8802-3 ANSI/IEEE 802.3 standards documents in the following sections: [7.5.2.5], [8.3.2.2], [10.4.2.2].

10 Mbps Ethernet uses Manchester encoding to transmit data across the AUI. It is a binary signaling mechanism that combines both the data and clocking into bit symbols. Each bit symbol is divided into two halves, with the second half containing the binary inverse of the first half. A transition always occurs in the middle of each bit symbol or bit period. During the first half of the bit symbol, the encoded signal is the logical complement of the bit value being encoded. During the second half of the bit

symbol, the encoded signal is the actual value of the bit being encoded. Since 10 Mbps Ethernet has no clocking mechanism between stations, the mid-bit transition provides the clocking necessary to allow the receiver to align its clock with each bit stream.

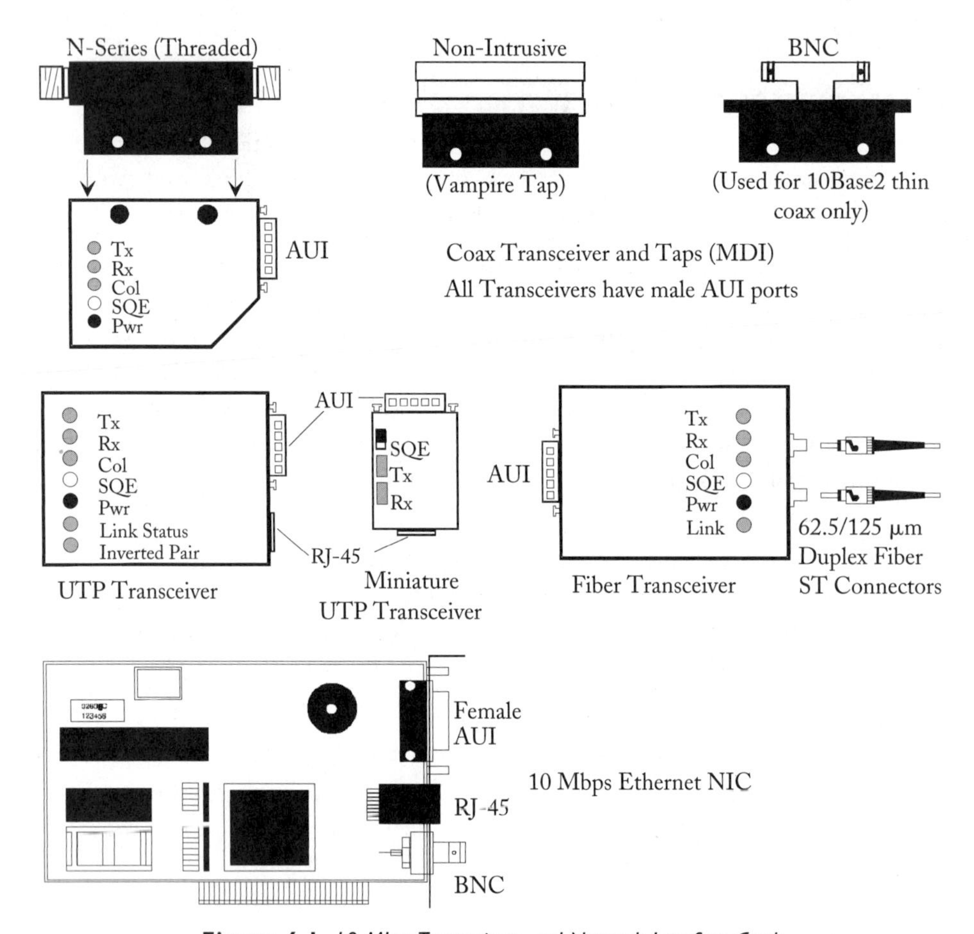

Figure 6.1 *10 Mbps Transceivers and Network Interface Cards*

TABLE 6.1 IEEE 802.3/ETHERNET TRANSCEIVER FUNCTIONS

Primary Functions	***Secondary Functions***
Transmit Data	Jabber Protection
Receive Data	Electrical Isolation
Collision Detection	Signal Quality Error Test (ISO/IEEE only)

Additional information may be found in the ISO/IEC 8802-3 ANSI/IEEE 802.3 standards documents in the following section: [7.3.1.1] (and previous).

Data signals are transmitted onto thick or thin coaxial medium at 0 volts to -2.05 volts (0mA to -82mA). Data signals are transmitted onto UTP medium using differential output voltage (i.e., equal negative and positive pulses). The peak differential voltage used on UTP media is between 2.2 volts and 2.8 volts (see Figure 6.2).

Collisions, Fragments, and Jams

Collisions are the result of two or more transceivers transmitting simultaneously. Receive mode Collision Detection results when two or more transceivers transmit on the same coaxial segment, causing the Collision Detect (CD) voltage threshold to be exceeded. This is often referred to as a *local collision*.

The voltage threshold for thick coax is between -1.492 V to -1.629 V. [8.3.1.5] The voltage threshold for thin coax is between -1.404 V to -1.581 V. [10.4.1.5] The variance in voltage is due to minor electrical differences between the two media. Modular transceivers that can support both types of coaxial media are either set to accommodate the lowest common denominator or provide a media selection jumper. Collision events are detected on UTP or fiber Link Segments when a transceiver senses activity on both transmit and receive circuits simultaneously. [14.2.1.4]

Transceivers enforce collision events by adding a 32-bit *Jam* to the transmission that caused the collision. The Jam signal may be any random string of bits and must not use the SFD bit pattern. [4.2.3.2.4], [4.4.2.1] Jam signals are also used to extend and propagate collisions across repeater-connected Ethernet segments. [9.5.6.2], [9.6.4] More details on repeaters will be provided later.

Collisions are normal occurrences in all (non-switched) IEEE 802.3/Ethernet environments. So-called packet fragments, runts, misaligned frames, and/or bad CRC values (defined below) are often the result of a collision. I take exception to the term *packet* because 802.3/Ethernet generates frames, not packets. These two terms are often used synonymously, which can sometimes lead to confusion. And contrary to what appears to be common knowledge, there are no consistently valid statistics or acceptable ranges for collision events or network utilization, nor are they measured consistently by various vendors' network monitoring equipment. Collisions are no measure of network performance; however, a collision LED on a hub or transceiver that remains illuminated most of the time is definitely indicative of a problem.

Packet fragments are actually incomplete Ethernet frames, or frames with intact headers and perhaps some data but a corrupt trailer. Runts are illegally short frames that are usually intact otherwise. Misaligned frames do not end on an even octet; they are missing a bit and are therefore not divisible by 8. A bad CRC indicates a loss of frame integrity during transmission; the result of the receiver's CRC calculation does not match the Frame Check Sequence (FCS) value in the frame's trailer.

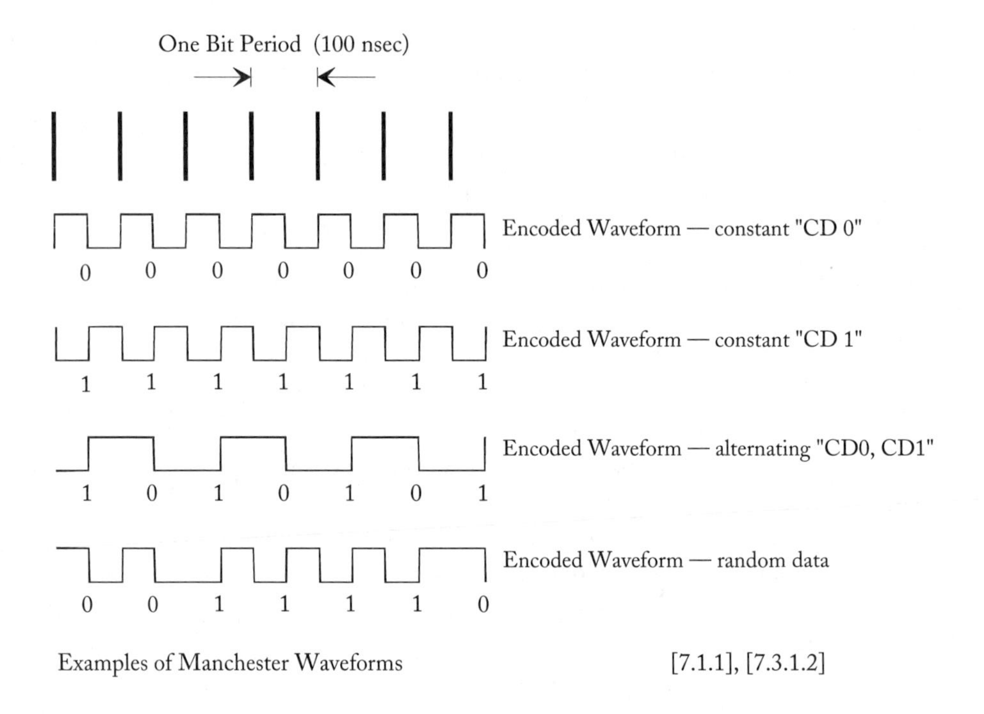

Examples of Manchester Waveforms [7.1.1], [7.3.1.2]

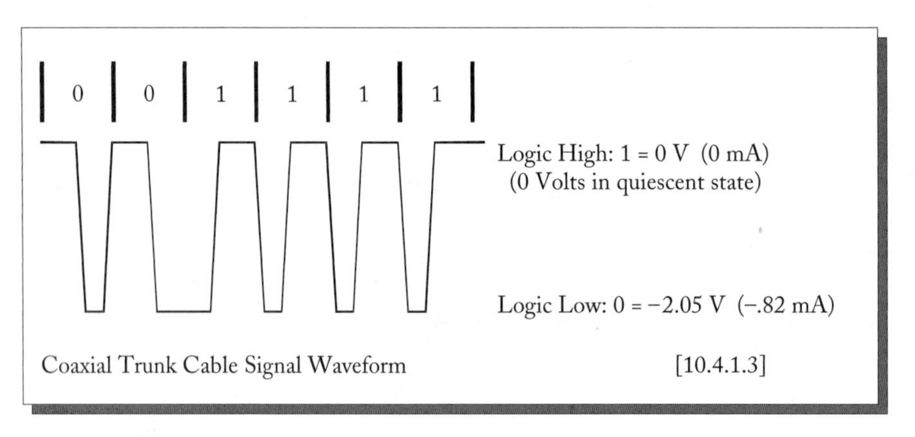

Coaxial Trunk Cable Signal Waveform [10.4.1.3]

NOTES:

CD = Clocked Data 1 and 0

Voltages given are nominal, for a single transmitter.

Rise and fall time is 25 ns nominal at 10 Mbps rate.

Voltages are measured on terminated coaxial cable adjacent to transmitting MAU.

Figure 6.2 *Manchester Encoding*

Jabber Protection

It is possible for a network interface card or transceiver to fail only partially. These components rarely fail, but when they do it is sometimes a slow, subtle death that generates intermittent network problems. One such fault condition causes the device to randomly transmit for extended periods that exceed the maximum transmission time permitted on an 802.3/Ethernet LAN. Transceivers include a feature called *Jabber Protection* to prevent this condition from disrupting the entire network.

Jabber Protection is a self-interrupt capability supported by IEEE 802.3–compliant transceivers that constantly monitors the AUI transmit lines (Data Out—DO) for excessively long transmissions.

The longest legal 10 Mbps Ethernet transmission has a duration of 1.2 milliseconds (or 1518 bytes). The IEEE 802.3 transceiver may opt to jabber-protect if the transmission exceeds 20 milliseconds, and it must jabber-protect by 150 milliseconds. This is accomplished by disabling DO, thereby silencing the transceiver. Some transceivers must be manually reset via a power cycle (off/on) while others will automatically reset after 0.5 seconds of silence on the DO leads. [8.2.1.5]

InterFrame Gap

All Ethernet frames must be separated by a minimum interframe spacing time called the InterFrame Gap (IFG). The minimum IFG specified by the IEEE 802.3 standard is 9.6 microseconds (about 12 Byte times at 10 Mbps). [4.2.3.2.1], [4.4.2.1] (See Figure 6.3.)

The purpose of the IFG is to allow enough time for the transmitting Ethernet adapter (or NIC) to cycle from transmit mode to receive mode. This delay was added

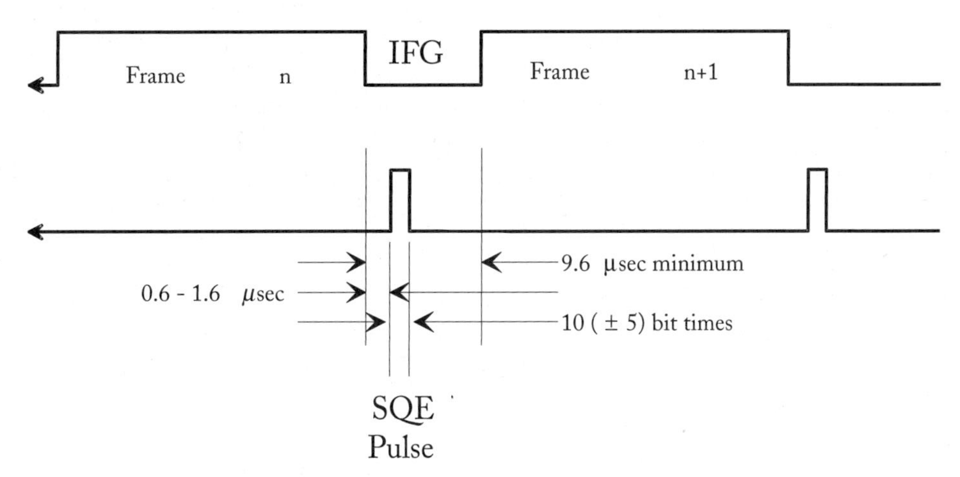

Figure 6.3 *InterFrame Gap and SQE*

due to a concern that the transmitting adapter may not switch from transmit mode to receive mode fast enough to read the header of a subsequent frame. If that frame was addressed to the station in question, it might be missed because the receiver was not ready. Technical advancements over the years have all but eliminated this problem, and some Ethernet adapter vendors have tweaked the performance of their cards by shortening the IFG, thereby providing better throughput than their competition. Such tweaking can sometimes result in a failure to interoperate with other vendors' cards on the same LAN.

IEEE 802.3 also specifies a Signal Quality Error (SQE) test. The original IEEE 802.3 committee was concerned that some mechanism should be implemented which would allow the host device (or DTE) to test the Collision Detection circuitry between the transceiver and the DTE via the AUI. To this end, an extra pair of conductors is included within the AUI cable as a Control Out (CO) pair. This CO pair allows the DTE to request the transceiver to assert SQE. This pair and function is optional and is not widely implemented. [7.5.2.3]

In addition, the IEEE specified that the transceiver should spontaneously assert the SQE message (pulse or burst) on the Control In (CI) pair of the AUI cable .6μs to 1.6μs after the end of *each* transmitted packet. The DTE will listen for the SQE message for 4.0μs to 8.0μs. The SQE message should be asserted for 10 (± 5) bit periods. The SQE test message occurs during the IFG and does not affect the LAN medium, interfere with data, or consume bandwidth. Most transceiver vendors provide for user-selectable SQE via an internal or external jumper or switch. [7.2.1.2.3], [7.2.4.6]

The worst-case variability of transmission elements in the network, plus some of the signal reconstruction facilities required in the repeater specification, combine in such a way that the gap between two packets traveling across the network may be reduced below the minimum (IFG). Interpacket gap shrinkage is a result of technological advances that have produced faster, more efficient network interface cards, and newer repeaters with lower latency and the ability to replace lost preamble bits (preamble extension). This phenomenon can be avoided by limiting the number of repeaters between any two DTEs and by adhering to the IEEE specifications for multi-segment 10 Mbps baseband networks (discussed later). [13.1], [13.4.2], [14.6], [A1.5.3]

SQE, Heartbeat, and Jam

As detailed above, SQE test is a function of IEEE 802.3–compliant transceivers that follows the transmission of every frame. The transceiver sends the SQE message to the DTE via the AUI CO line after each SQE test and when collisions are detected. This is the same pair of conductors in the AUI used to indicate collision presence. [8.2.2.2.4], [10.3.2.2.4]

The SQE test, sometimes referred to as heartbeat, *must* be disabled on all transceivers connected to repeaters. Repeaters (discussed later) implement only the Physical layer

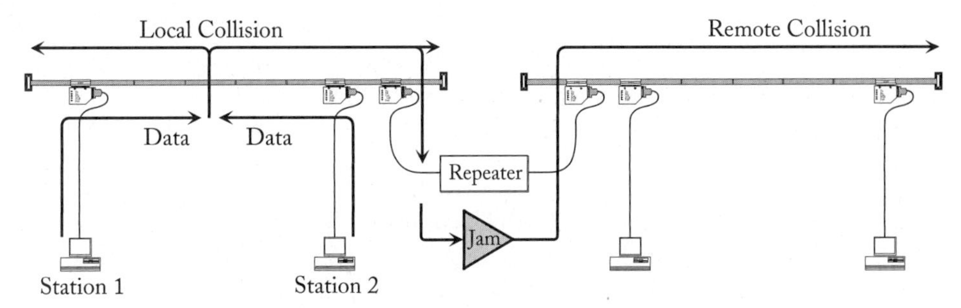

Figure 6.4 *Collision, Jam, and Repeaters*

component of the 802.3 specifications and misinterpret all SQE messages as actual collisions. Upon receiving the SQE message, a repeater reacts as if it sensed an actual collision and propagates a real collision event by transmitting at least 96 bits of Jam signal on all other repeater ports (but not on the port of origin). (See Figure 6.4.)

A Jam produces a packet fragment and is used to enforce a collision event across all repeater-connected segments. By misinterpreting the SQE message, the repeater generates artificial collisions as it transmits each frame. The Jam is not transmitted onto the segment of origin (i.e., the port with the transceiver that asserted SQE), but it does affect all other segments of that repeater. In other words, if SQE testing is enabled on a transceiver connected to a repeater, it will cause excessive collisions to appear on all other Ethernet segments—except the segment where the problem is located.

In small networks with little traffic, this problem may go unnoticed. As the network traffic increases, the number of collisions escalates until the network grinds to a halt. During this phenomenon, repeater ports may repeatedly auto-partition and reconnect, users may complain of sporadically poor performance, or application sessions may be randomly dropped. These problems are usually attributed to too many stations or too little bandwidth rather than a misconfigured transceiver.

Electrical Isolation

IEEE 802.3/Ethernet transceivers must provide electrical isolation between the AUI cable and the coaxial trunk cable. The isolation impedance measured between each conductor (including shield) of the AUI cable and either the center conductor or shield of the coaxial cable shall be greater than 250k Ohms at 60 Hz and not greater than 15 Ohms between 3 MHz and 30 MHz. This electrical isolation must withstand 250 volts AC or 500 volts AC rms for one minute (for 10Base5 and 10Base2, respectively). This prevents the occurrence of a ground loop, which can disrupt data flow and pose an electrical hazard to personnel and equipment. [8.3.2.1], [10.4.2.1]

The transceiver must be connected *as closely as possible* to the network medium via the transceiver tap, referred to by the IEEE as the Medium Dependent Interface (MDI).

The 10Base5 transceiver and *vampire* tap or N-series tap are normally considered to be one assembly. The connection between the coax and the transceiver must be within 30 mm (1.81"). The connection between the 10Base2 transceiver and BNC "T" must be within 40mm (1.575"). [8.5.3], [10.6.3]

Grounding

10Base5: The shield conductor of each thick coaxial cable segment must be grounded at one point and must not make electrical or ground contact on other objects such as building structural metal, ducting, plumbing fixtures, or other unintended conductors. Insulating sleeves or boots may be used to cover any coaxial connectors used to join cable sections and terminators. A ground lug with a current rating of at least 1500 ampacity must be provided on one of the two terminators or on one extension connector used along the length of each cable segment. The shield conductor of the AUI cable shall be connected to the chassis (or earth reference) of the DTE. [8.6.3.2], [8.7.2.2], [8.7.2.3]

10Base2: Thin coax Ethernet transceivers that are embedded on the NIC must provide a static discharge path to the DTE/station ground. This is done via a 1 M Ohm, 0.25-watt resistor that has a rating of at least 750 volts DC. Earth grounding of thin coax is optional. [10.7.2.3], [10.7.2.4]

The 10Base2 BNC "T" connector *must not* touch the metal chassis of the DTE/station or other metal connectors that may be near the NIC. If the BNC "T" contacts the chassis or another connector plug, it may provide a direct path to ground via the DTE. This can cause the network to behave erratically or to halt completely.

DIX Drop Cables and IEEE AUI Cables

Commonly referred to as an Ethernet drop cable or branch cable, the IEEE 802.3 term for the transceiver cable is Attachment Unit Interface. The AUI is the common denominator to which all 10 Mbps IEEE 802.3/Ethernet standard components must adhere. All devices gain access to the Ethernet media via a transceiver through the AUI, whether it is a physical cable or a service interface circuit embedded on a network interface card or adapter. [7.4.3] (See Figure 6.5.)

AUI cables consist of individually shielded twisted pairs with an overall braided metal shield. The special DB-15 connectors are male with locking posts on the DTE/station end, and female with an unreliable slide latch on the transceiver end. The cable pin-out is straight through—no crossed pairs. *Don't even think about using a "gender bender."* If a gender-changing adapter appears to be a requirement, you are attempting to do something that is technically wrong. The AUI cable provides power to the transceiver, as well as Collision Detection, and transmission and reception of data to the DTE.

The AUI cable specification permits variable lengths of up to 50 meters (164 feet). Although the IEEE does not specify the conductor gauge, the typical AUI cable uses 20–22 AWG conductors. An alternative to this heavy, bulky, and relatively expensive

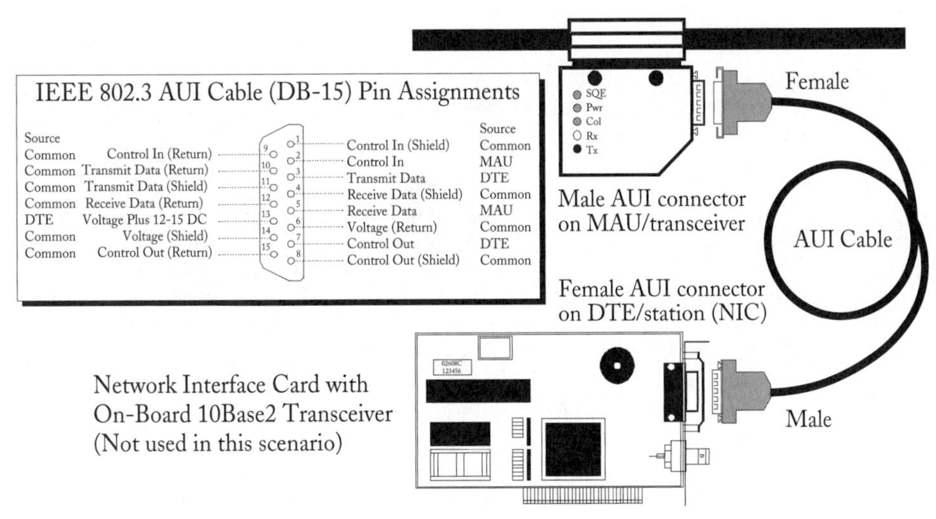

Figure 6.5 *IEEE Attachment Unit Interface and DIX Drop Cables*

cable (with poor connectors that collapse under the weight of the cable) is a 24–26 AWG version that most vendors will support to a maximum length of 10 feet (about 3 meters).

All AUI/Ethernet drop cables are *not* the same. The IEEE 802.3 AUI cable pin-out is different from the earlier DIX ESPEC1 and ESPEC2 Ethernet drop cable pin-out. A quick way to identify an 802.3 AUI cable is to look at the metal "D" shell of the DB-15 connectors. The IEEE specifies that the male "D" shell must have indentations along its two long sides. This is intended to provide a superior ground contact between the shells of the male and female connectors.

> **Note:** ESPEC v.1 drop cables and components are not compatible with ESPEC v.2, IEEE 802.3, or ISO 8802-3. The latter three can peacefully coexist in the same LAN.

Multiport Transceivers, DELNIs, and Fan-Outs

Multiport transceivers (MPT) provide a number of male AUI transceiver ports that may be used to interconnect that number of DTE/stations. They also provide one female AUI port for connection to an external transceiver attached to a coax, UTP, or fiber optic Ethernet network. With the advent of cheap 10Base-T NICs and hubs, these devices (along with coax-based LANs in general) have all but completely disappeared (see Figure 6.6).

MPTs were also called *fan-out units* because they can share a single network-attached transceiver among a number of stations. They may be used in any location where additional Ethernet devices require a connection to a network, but there is no space on the coax (or enough UTP, or fiber runs) for more transceivers, or when an alternative

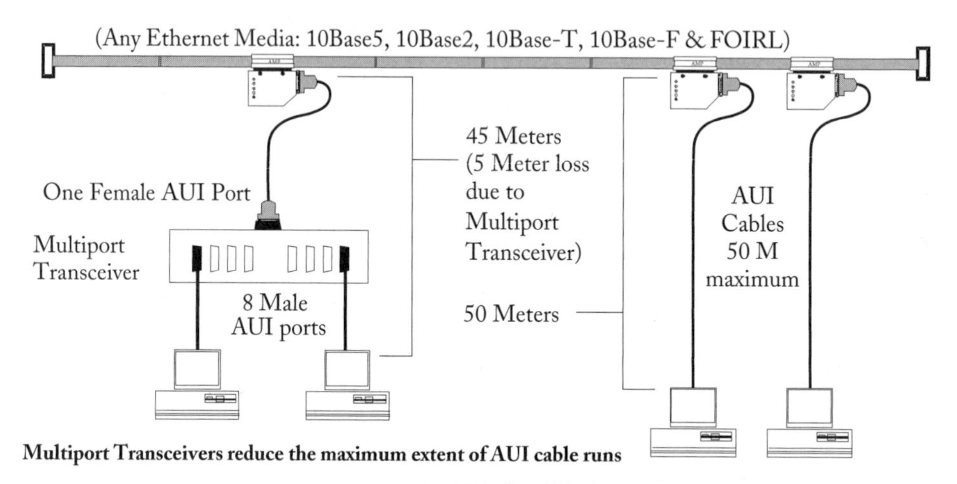

Figure 6.6 *Multiport Transceivers*

solution is inconvenient. MPTs require AC power, are relatively *dumb,* unmanageable devices, and are another potential point of failure.

DELNI is the name given to Digital Equipment's MPT. It supports eight male AUI ports and one female AUI port that can be set to the loopback mode. The term has gained such popularity that other vendors' MPTs are also referred to as DELNIs. (Digital was acquired by Compaq, which was acquired by HP.)

10BASE5

PHYSICAL MEDIA

Thick coax was the most common type of Ethernet media installed until the introduction of products that support Ethernet over unshielded twisted pair (UTP) and, to a lesser degree, fiber optic media. Thick coaxial Ethernet is also referred to as Standard 802.3 10Base5, N-series Ethernet, or Ethernet Trunk Cable (see Figure 6.7). Table 6.2 gives specifications for thick coax Ethernet.

TRANSCEIVERS AND TAPS

There are two basic forms of transceiver taps used on thick coaxial Ethernet. Both may be installed on a common cable. The two forms are N-series and "non-intrusive."

N-series, type N, or intrusive taps require cutting the coax and installing a threaded N-series connector on each end. Since the cable must be cut, transceiver installation

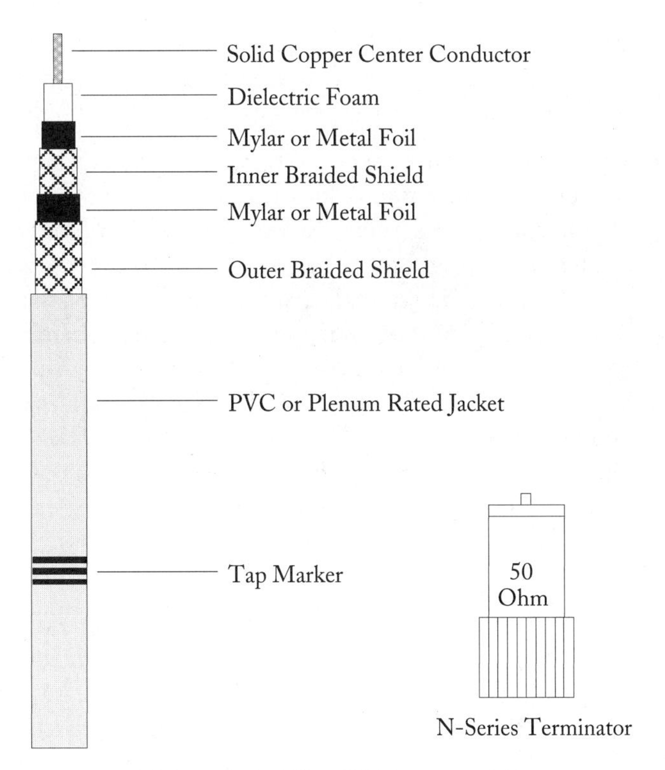

Figure 6.7 *10Base5 Thick Coaxial Cable*

TABLE 6.2 THICK COAX ETHERNET SPECIFICATIONS

Characteristic Impedance [8.4.1.1]	50 Ohms ±2 Ohms
Attenuation not to exceed [8.4.1.2]	8.5 dB/500m @ 10 Mhz (17 dB/km) 6.0 dB/500m @ 5 Mhz (12 dB/km)
Velocity of Propagation [8.4.1.3]	.77 c
Edge Jitter (untapped cable) [8.4.1.4]	< 8.0 ns/500m
Minimum Bend Radius [8.4.2.1.1(4)]	254 mm (10")
Termination (@ each end) [8.5.2.1]	50 Ohms ±1 % (1w power rating) Measured from 0–20 MHz
Typical Weight per 1000'	116 lbs

using N-series taps always forces a disruption of network service until the installation is completed. Since most thick coax LANs remain static once installed, the flexibility offered by the N-series tap is of little value (see Figure 6.8).

Some network engineers prefer the N-series to the non-intrusive vampire taps, claiming they provide a superior connection. Non-intrusive taps come in several variations and do not require the coax to be severed. Referred to as a vampire tap or bee-sting tap, the typical installation involves mounting the MDI—a metal clamp—onto the coaxial cable. Once the clamp is secured, a special drilling tool is inserted into a round opening in the transceiver attachment side of the clamp assembly. The drilling tool is operated manually and only requires a few turns to bore through the coaxial jacket, the two shields, and slightly into the white dielectric foam material that surrounds the solid copper center conductor. This procedure does NOT require great force and the drill bit should never touch the center conductor. Take care not to permit any cable shavings to remain in the freshly bored hole, as this may cause a short or a poor contact between the center conductor and ground. Also, moisture must never be introduced into the cable (see Figure 6.9).

The center tap is then screwed into the threaded hole in the clamp using a wrench that is usually provided on the other end of the drilling tool. The tap should be completely screwed into the clamp so that the metal point on the end of the tap makes a secure contact with the coaxial center conductor. Again, too much force can break the tip/point of the tap or damage the actual center conductor of the coaxial cable.

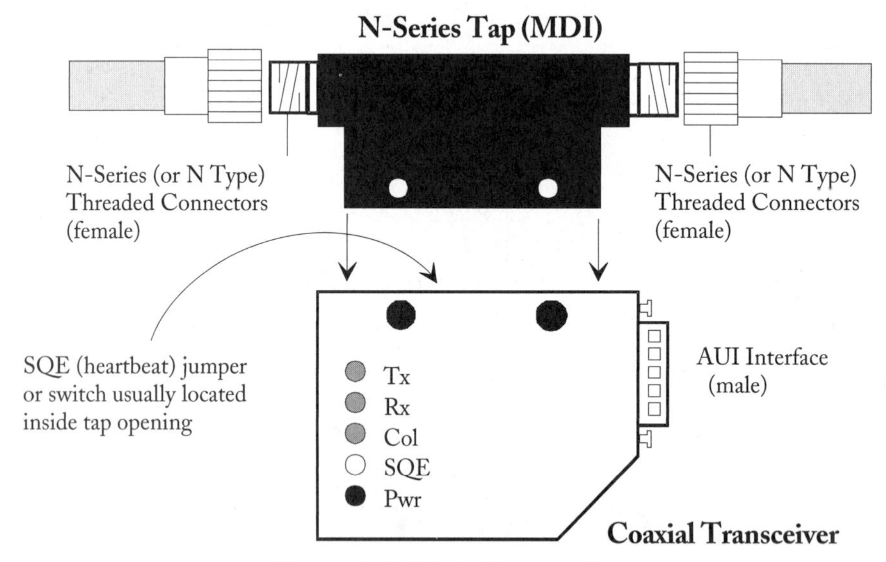

Figure 6.8 *N-Series Transceiver Tap*

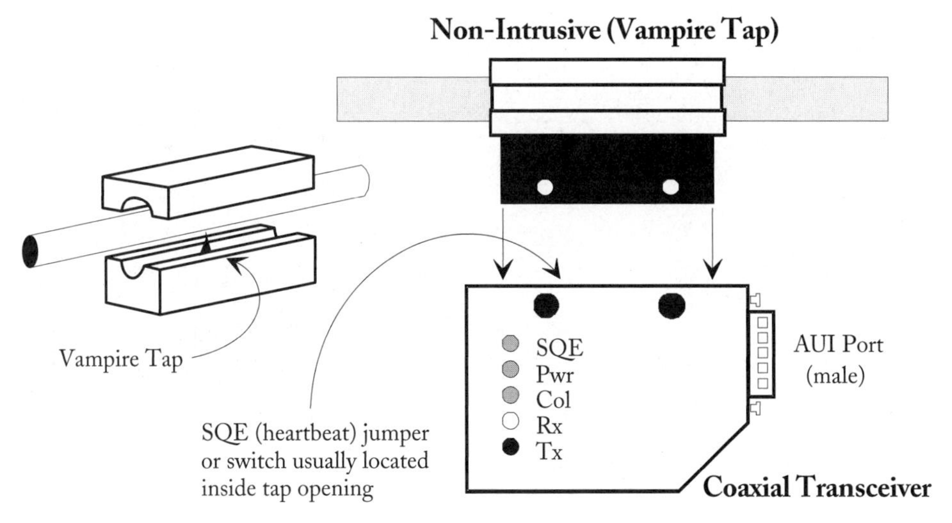

Figure 6.9 *Non-Intrusive Vampire Tap*

Later varieties of non-intrusive taps install much like the water supply line of an automatic icemaker. The clamp is securely mounted onto the coax, and the actual tap is screwed into the clamp—no drilling required. The tap itself is insulated and sturdy enough to puncture the coaxial jacket, displacing both shields and the white dielectric material to achieve a secure contact with the center conductor. Although non-intrusive tap installation is sometimes performed on active networks, it is *always* recommended that an active network be brought down gracefully and all users of the system be forewarned of scheduled network modifications or what may potentially turn into extensive downtime.

INSTALLATION RULES

10Base5 specifies a maximum segment length of 500 meters (1640 feet) with a total of 100 transceiver attachments per segment. Transceivers are spaced every 2.5 meters (8 feet) along the length of the cable. Each end of the coaxial segment is terminated with a 50-Ohm resistor, and the coax must be grounded at one point. [8.1.1.1], [8.4.2], [8.5.2.1] (See Figure 6.10.)

The term *trunk* is generally used in reference to a thick coaxial cable segment, which is terminated at each end with a 50 Ohm terminating resistor. The trunk/segment should be earth-grounded at one point only, either at the coaxial terminator or an inline extension connector (barrel). The ground lug should have a current rating of 1500 ampacity. Check with the local authorities regarding earth grounding and other electrical codes.

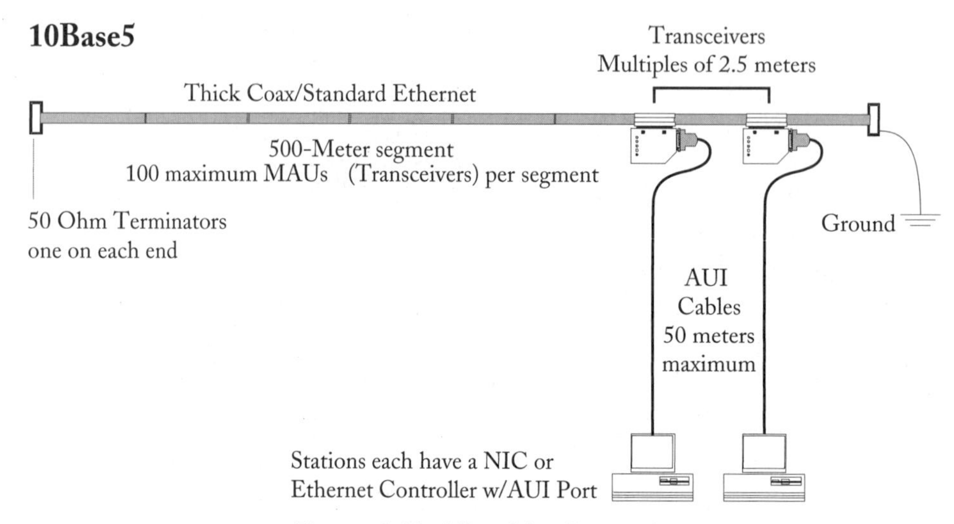

Figure 6.10 *10Base5 Installation Rules*

A *plenum* is an air-filled space, usually above the ceiling, that is used as an air-return duct for heating, ventilation, and air conditioning systems (HVAC). Most building electrical codes require the use of special fire retardant or plenum-rated cable if cabling is to be installed in such spaces. In addition, the coaxial metal shield must not make electrical contact with earth ground elsewhere (i.e., building structural metal, ducting, plumbing fixture, or other unintended conductor). [8.5.2.2], [8.6.2.3]

The maximum allowable thick coaxial cable segment length is 500 meters (1640 feet). The media is to be constructed of a brightly colored outer jacket (usually yellow) to avoid being mistaken for power mains. The jacket markings, also referred to as tap markers, are annular rings that must contrast with the background jacket color (usually black). The tap markers are spaced at 2.5 meters (± 5 cm) regularly along the entire length of the cable. Transceivers may be placed only on these tap markers (e.g. 2.5 m, 5 m, 7.5 m, 10 m). This is done to ensure nonalignment on fractional wavelength boundaries. If wavelength boundary alignment occurs, undesirable network-disrupting events such as standing waves may form. [8.4.2.2], [8.6.2.2]

10BASE2

PHYSICAL MEDIA

Table 6.3 gives specifications for thin coax Ethernet.

It is recommended that *all* metal parts of the cable connectors be insulated to prevent touching any part of the building metalwork which may be at ground potential, or unintended conductors. The IEEE recommends rubber or plastic boots and sleeves (see Figure 6.11).

TABLE 6.3 THIN COAX SPECIFICATIONS

Characteristic Impedance [10.5.1.1]	50 Ohms ±2 Ohms
Attenuation (185 meters) [10.5.1.2]	8.5 dB/185m @ 10 Mhz 6.0 dB/185m @ 5 Mhz
Velocity of Propagation [10.5.3.1]	.65 c
Edge Jitter (untapped cable) [10.5.1.4]	< 8.0 ns/185m
Minimum Bend Radius [10.5.2.1(2)]	5 cm (2")
Termination (@ each end) [10.6.2.1]	50 Ohms ±1 % (.5 watt power rating) Measured from 0–20 MHz
Typical Weight per 1000'	22 lbs

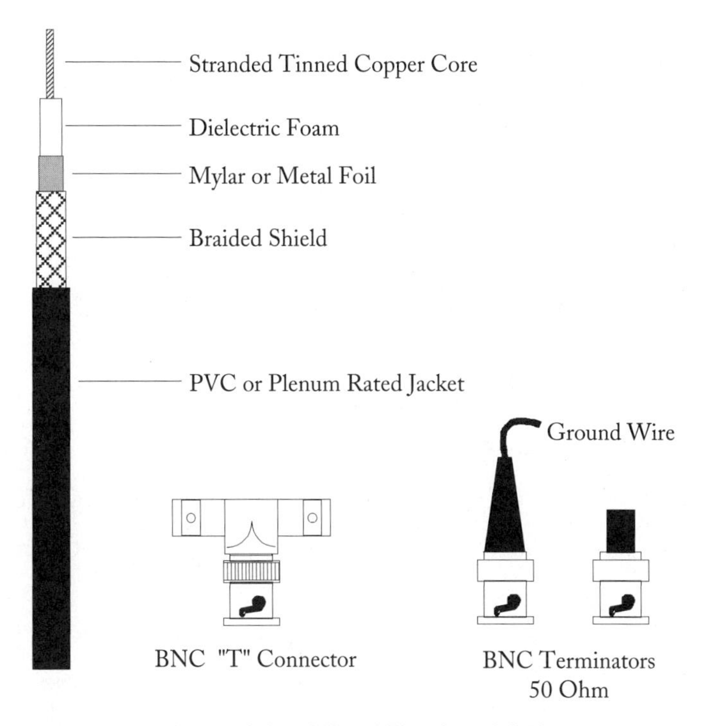

Figure 6.11 *10Base2 Thin Coaxial Cable*

Thin coaxial Ethernet is also referred to as ThinNet, CheaperNet, or ThinWire. There are many terms or product numbers given to thin coaxial cable that is sold for use in Ethernet networks. These numeric terms are not used consistently from vendor to vendor, nor are they guaranteed to meet all of the correct electrical characteristics for use as Ethernet media. Look for cable identified as RG *spec* rather than RG *type*. RG Type cables are frequently 53 Ohms rather than the specified 50 Ohms.

The only way to be certain of obtaining the correct media is to request that the product be certified for use with IEEE 802.3 10Base2 Ethernet. Do not accept cheap substitutions, or claims that one cable is equal to another. Ask for documented proof. Many cable installers do not like the stranded-center coax because it is more difficult to install connectors on than the solid conductor variety. Solid-center coax will actually support Ethernet fairly well in very small LANs. It fails when the technology is pushed near the limit of 185 meters (and/or 30 transceivers). Many installers and suppliers are not at all aware of these differences or the problems they can cause. The 802.3 10Base2 standard requires the center conductor to be stranded, tinned copper with an overall diameter of 0.89 mm ± 0.05 mm. [10.5.2.1.2]

> **Note:** Avoid implementing coax segment lengths longer than the IEEE standard specifications even if the vendor endorses such installations using their products. These implementations are non-standard and will usually create havoc in mixed-vendor environments. (These ideas are great for marketing—terrible for networking.)

IEEE approved cables for use with 802.3 10Base2 Ethernet are:

802.3 10Base2	50 Ohms, Stranded Tinned Copper Core
RG-58 a/u	50 Ohms, Stranded Tinned Copper Core
RG-58 c/u	50 Ohms, Stranded Tinned Copper Core

The following media types are inappropriate:

RG-58	50–54 Ohms, Solid Center Core
RG-58 u	50–54 Ohms, Solid Center Core
RG-59	75 Ohms, Solid Center Core (CCTV, PC Net)
RG-62	93 Ohms, Solid Center Core (IBM 3270, Arcnet)

TRANSCEIVERS, TAPS, AND TEES

All connections to thin coaxial Ethernet are accomplished via a BNC "T" connector or tap. The BNC-T is installed in-line, as part of the linear coaxial bus. As with thick coaxial taps, the transceiver is connected directly to the coax. Instead of mounting a large metal clamp on the coax (as with thick coax installations), the thin coax MDI uses a simple BNC-T connector. The male connector of the BNC-T must attach directly to the transceiver or the female BNC connector on each NIC.

The physical configuration of thin coax requires two cables between the coaxial wall jack and each station's NIC. Single lengths of coax between the BNC-T and the NIC connector are called "spurs" or "stubs" and are *not* permitted (see Figure 6.12).

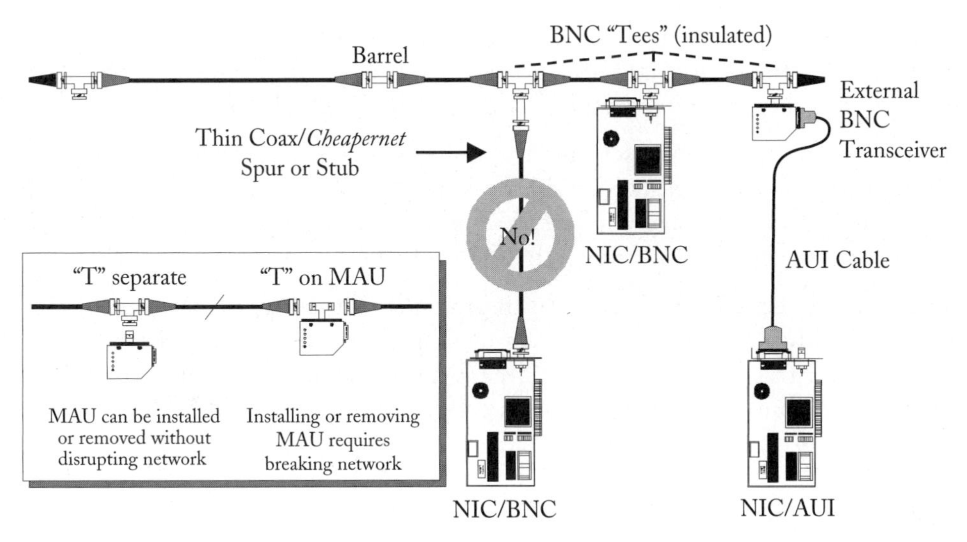

Figure 6.12 *Transceivers, Taps, and Tees*

10Base2

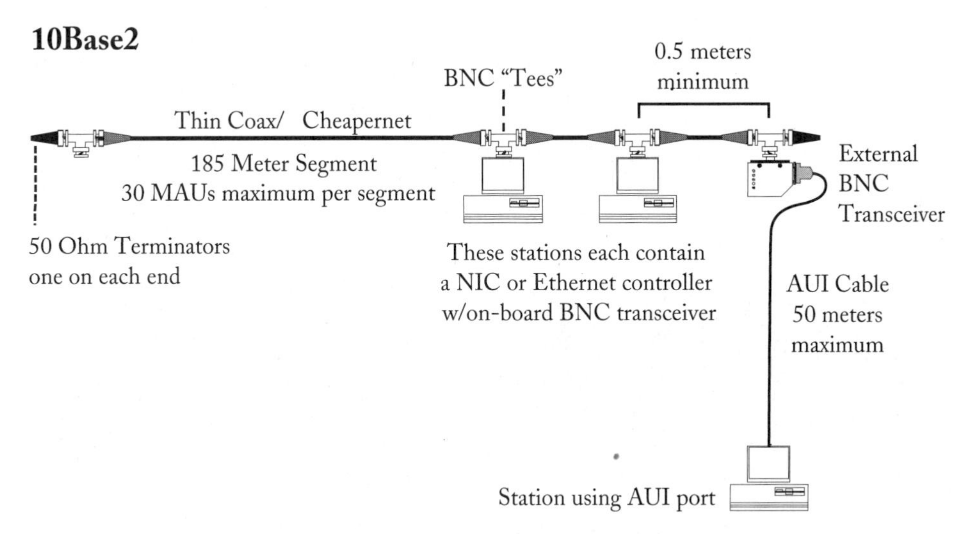

Figure 6.13 *10Base2 Installation Rules*

INSTALLATION RULES

10Base2 specifies a maximum segment length of 185 meters (606.8') with a total of 30 transceiver attachments per segment. Like 10Base5, each end of the coaxial segment is terminated with a 50-Ohm resistor. [10.1.1.1], [10.6.2.1] (See Figure 6.13.)

External BNC transceivers may also be used on thin coaxial Ethernet to support devices that have no on-board thin coax transceiver and BNC connector. These transceivers may be identical to those used on 10Base5, only a BNC tap is employed in place of either the N-series or non-intrusive coaxial taps. 10Base2 transceivers may be placed no closer together than .5 meters (about 20 inches). [10.6], [10.6.3]

Alternative thin coax cabling schemes that employ a convenient plug-in jack wall fixture and special dual-coax station cable look nice but have always performed poorly. Such cabling systems usually reduce the number of 10Base2 transceivers/stations supported on each segment from 30 to only 10 or 11.

Rarely do thin coax LANs require earth grounding (it is at least rare to find them grounded). Most Ethernet NICs with on-board 10Base2 transceivers provide a capacitive shunt to the chassis ground of the PC. Grounding is recommended in installations where thin coax is used only with external transceivers. [10.7.2.3]

10BASE-T

PHYSICAL MEDIA

Unlike other standards that define a maximum distance, IEEE 802.3 10Base-T stipulates that the UTP medium and components must *operate over 0 meters to at least 100 meters* (328 feet) without the use of a repeater. The standard supports point-to-point interconnections, as well as a star-wired topology using multiport repeaters, oth-

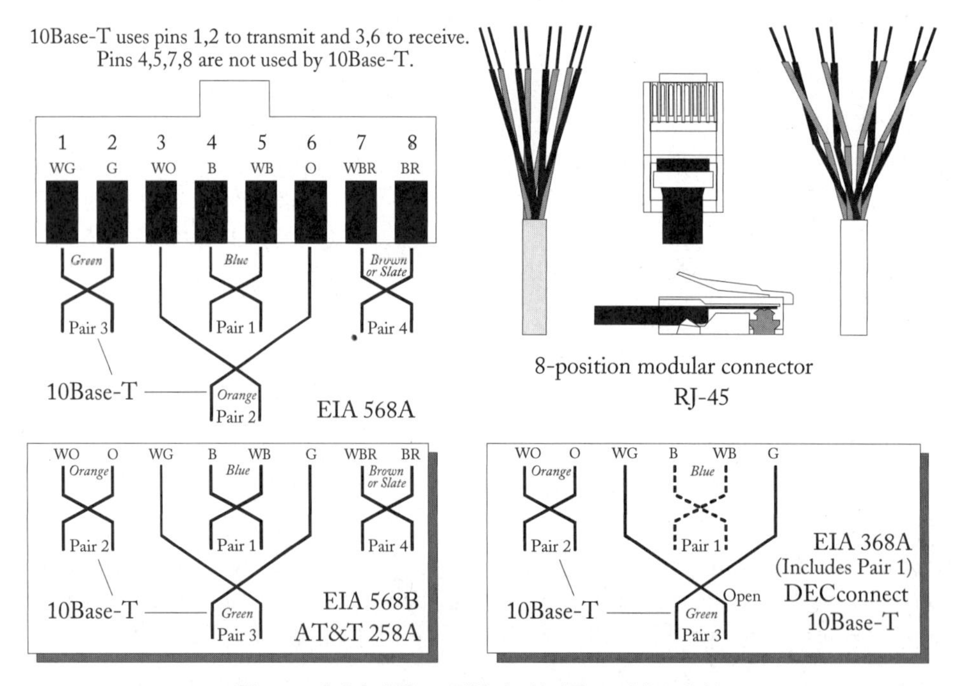

Figure 6.14 *10Base-T Unshielded Twisted Pair Cable*

erwise known as hubs. Repeater/hubs are integral to all 10Base-T networks with more than two DTE/stations. And contrary to popular belief, 10Base-T was engineered to support the electrical characteristics of ordinary (26 AWG – 22 AWG) UTP telephone wire, not TIA/EIA Category 3 UTP. This is because the standards for Category 3 UTP were defined *after* the 10Base-T standard was released. 10Base-T requires media that meets or exceeds the following electrical characteristics (see Figure 6.14).

Table 6.4 gives specifications for UTP.

The Insertion Loss of a simplex Link Segment must be no more than 11.5 dB at all frequencies between 5 and 10 MHz. This consists of the attenuation of the twisted pairs, connector losses, and reflection losses due to impedance mismatches between the various components of the simplex Link Segment. Multi-pair Poly Vinyl Chloride (PVC)–insulated .5 mm [24 AWG] cable typically exhibits an attenuation of 8–10 dB/100 m at 20° C. [14.4.2.1]

No more than ±5.0 ns of medium timing jitter may be introduced to a test signal by a simplex Link Segment. Inter-symbol interference and reflections due to impedance mismatches between tandem twisted pairs of a twisted pair Link Segment and effects

TABLE 6.4 UTP SPECIFICATIONS

Differential Characteristic Impedance [14.4.2.2] (or)	85–111 Ohms @ 5.0–10MHz 100 ±15 Ohms @ 1–16 MHz
Attenuation (Max. Insertion Loss) [14.4.2.1]	< 11.5 dB
	@ 5.0–10 MHz
Minimum Velocity of Propagation [14.4.2.4]	.585 c (5.7 ns/m)
Maximum Medium Timing Jitter [14.4.2.3]	± 5.0 ns
Near End Crosstalk (NEXT) [14.4.3.2.1]* (Loss between any two pairs in a four-pair cable)	5 MHz > 30.5 10 MHz > 26.0 10 MManchester > 20.5
Multiple Disturber NEXT (MDNEXT) Loss [14.4.3.2]* (Loss between Tx and RX pairs in a binder group)	5 MHz > 27.5 10 MHz > 23.0 10 MManchester > 20.5
Impulse Noise Rate [14.4.4.1]	Less than 0.2 per second occurrences of bursts of >264 mVolts. [or < 180 hits/15 minutes]
Typical Bend Radius	8 × Outside Diameter
Typical Weight per 1000' (4 pair DIW 24)	18.6 lbs

of connection devices can introduce jitter to the CD1 and CD0 signals (data stream) received on the RD circuit. The test signal requires peak amplitude of 3.0V and 10–90% rise and fall times of 12 ns. Test signal content should be a Manchester-encoded pseudo-random sequence with a minimum repetition of 511 bits. Branches off a twisted pair (often referred to as "bridge taps" or "stubs") will generally cause excessive jitter that can disrupt data transmissions and should therefore be avoided. [14.4.2.3]

TRANSCEIVERS, CONNECTORS, AND HUBS

The 10Base-T standard specifies use of a modular 8-position connector (commonly referred to as the RJ-45). Most RJ-45 connectors are designed to accommodate a jacketed, round, 4-pair solid conductor cable. The strain-relief for the cable is provided by part of the RJ-45 connector that acts as a wedge against the outer jacket of the 4-pair cable. The wedge is pressed and locked tightly against the cable jacket when the connector is crimped into place.

Do *not* attempt to save money by using 2-pair cable with RJ-45 connectors. Although technically adequate and a bit cheaper, the RJ-45 connector will not securely fasten onto the 2-pair cable. All cable stress will be transferred directly to the conductors and will eventually cause the connector to fail. Some cable vendors will attempt to resolve this problem by applying a short piece of heat-shrink tubing to each end of the 2-pair cable to add to the cable jacket's diameter. This kludge does not last. Building cables is a craft that requires a great deal of practice to produce consistently reliable cables. Rather than making patch cables on-site, purchasing pre-made patch cable assemblies is highly recommended.

The 10Base-T cable that connects the station/DTE to the repeater/hub is called a Link Segment. Link Segments are point-to-point only and require a transceiver at each end—one at the station end and the other at the hub or concentrator. Each port on a 10Base-T hub has a transceiver—i.e., a 24-port 10Base-T hub consists of 24 transceivers connected to a common repeater circuit such that all 24 ports are a single *repeater hop* away from each other (see Figure 6.15).

Somewhere along the way the transmit pair and the receive pair of each Link Segment must be crossed. The 10Base-T standard specifies that this is always done at the transceivers in the hub and that each port must be identified as the Medium Independent Interface–Crossover, or MDI-X. The station/DTE transceivers are referred to as simply MDI, and are pinned out straight through (see Figure 6.16).

A Null Link Segment cable that performs the transmit and receive pair crossover can be used to directly connect two stations without using a hub. Null Link Segments are sometimes used to cascade one hub off another; however, most hubs provide a separate (or switched) MDI up-link port for this purpose (see Figure 6.17).

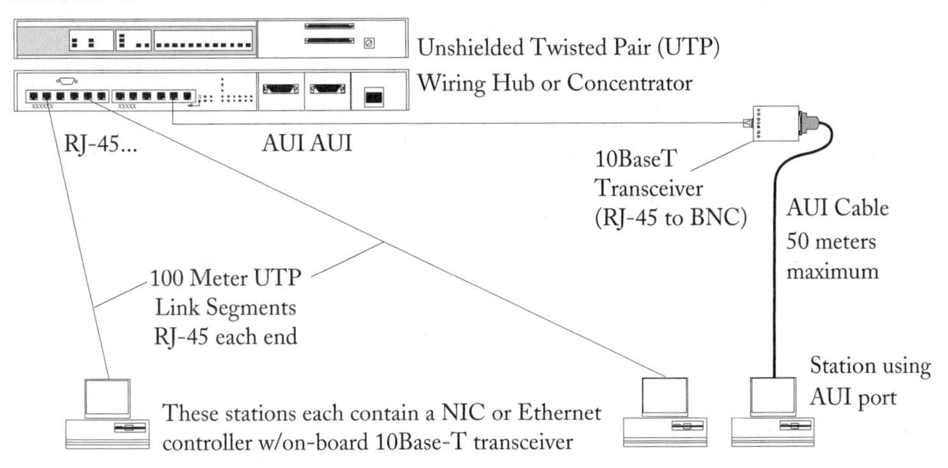

Figure 6.15 *Transceivers and Hubs*

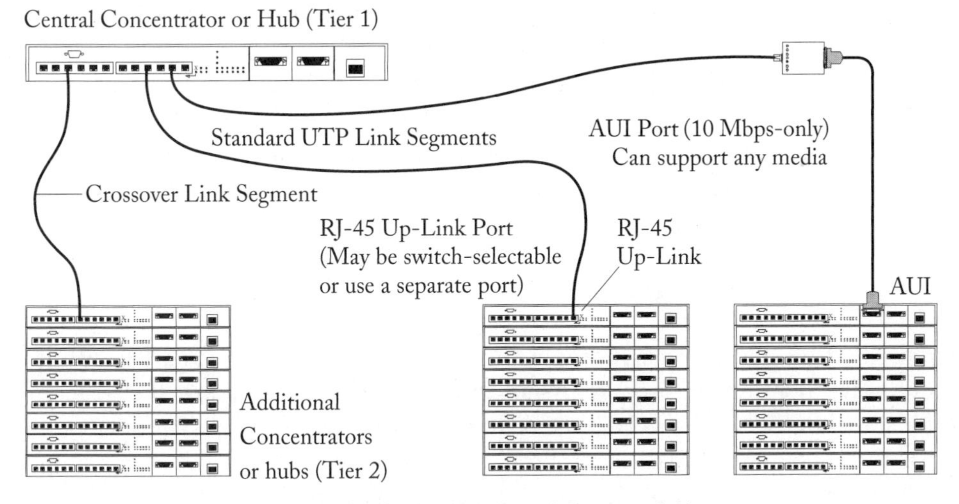

Figure 6.16 *Hub Uplinks and Crossover Cables*

The number of telephone punch-down blocks a single Link Segment or station run may traverse is generally limited by most vendors to five. The ANSI/EIA/TIA is more stringent, in most cases limiting a Link Segment to two punch-down blocks: hub to patch panel, patch panel to office wall jack, office wall jack to end station. Virtually all patch panels and wall jacks provide an integrated punch-down block for cable termination.

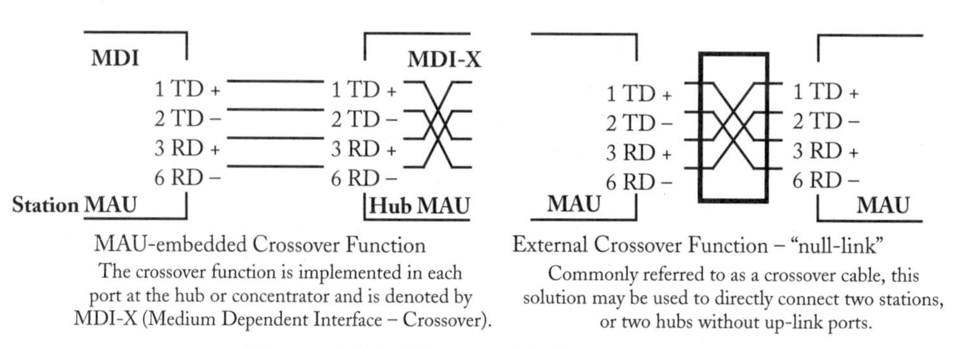

Figure 6.17 *MDI and MDI-X Crossover Function*

Avoid using wall jacks with screw terminals, as they are not certified for data. Also avoid using the Universal Service Order Code (USOC) wiring. Because of the split pairs (of the connector pin-out specification), the USOC scheme is among the poorest possible wiring configurations for data transmission.

10Base-T transceivers signal their presence by issuing a link pulse signal over their own transmit pair. When the transceiver of the hub port receives the link pulse from the station/DTE connected to the Link Segment attached to that port, it will illuminate that port's link status LED. At the same time, when the transceiver on the station's NIC receives the link pulse from the hub's transceiver, it will illuminate its link status LED. The link status indicator is not provided by all NICs, although they are supposed to. Note that just because you get a pair of green lights on each end of a given Link Segment it does not guarantee that data can be successfully transmitted! It merely means that you have continuity.

INSTALLATION RULES

Unshielded twisted pair has become the most popular LAN media because it is inexpensive, light, small, supports a star-wired topology, and employs the structured cabling techniques used in telephone systems. A structured cabling system refers to a cable plant design philosophy that is both manageable and serviceable. Structured cable plants are star-wired to provide simple and economical moves, adds, and changes, as well as superior maintenance, monitoring, and management capabilities unavailable with most other cabling schemes. Structured cabling systems also provide clear definitions for the design and installation practices as well as the media used.

For years the most common UTP cable used was AT&T DIW24; two to four twisted pairs, approximately two twists per foot, solid conductors (22–26 AWG), and no shielding. The 10Base-T structured cable plant may be implemented using typical telephone punch-down blocks, patch panels, and wall outlets. The punch-down blocks may be any type such as the ubiquitous "66" block, AT&T Premises Distribution

System (PDS) 110, and others. Most styles of punch-down blocks (connecting or terminating blocks) are acceptable for use with 10Base-T, but avoid using bridge taps. Bridge taps are simple metal clips used to form a circuit path between adjacent punch-down terminals (see Figure 6.18).

The popularity of DIW24 cabling eventually gave way to TIA/EIA Category 3 UTP (Cat3) when it became available. Cat3 provided somewhat better electrical characteristics and greater product consistency, as well as installation specifications. The EIA/TIA specified a 100-meter channel. The channel is the entire UTP Link Segment from hub to station/DTE and allocates 90 meters (295 feet) to premises cabling, 5 meters (16 feet) for patch cords and another 5 meters (16 feet) for station cords, for a total of 100 meters (16 feet). (See Figure 6.19.)

One of the most critical aspects of UTP cabling is the maintaining of correct pairing throughout the cable plant. Simply checking for continuity is not adequate for 10Base-T cable plant certification. The two conductors that make up each pair must remain as a pair end-to-end through the entire channel. Do not split pairs! Split pairing occurs when individual conductors are separated from their pair-mate, joined with another mismatched pair-mate, and then rejoined with their original pair-mate. As far-fetched as this may seem, it has been known to happen.

In addition, Silver Satin, which is used extensively in telephony installations, is *not* acceptable for use in 10Base-T cable plants. Digital Equipment Corporation's DECconnect flat, gray patch cable is also unacceptable. In most cases the connectors will fit and you may even achieve link status, but the poor electrical characteristics of these cables will generate high bit-error rates, which degrade performance.

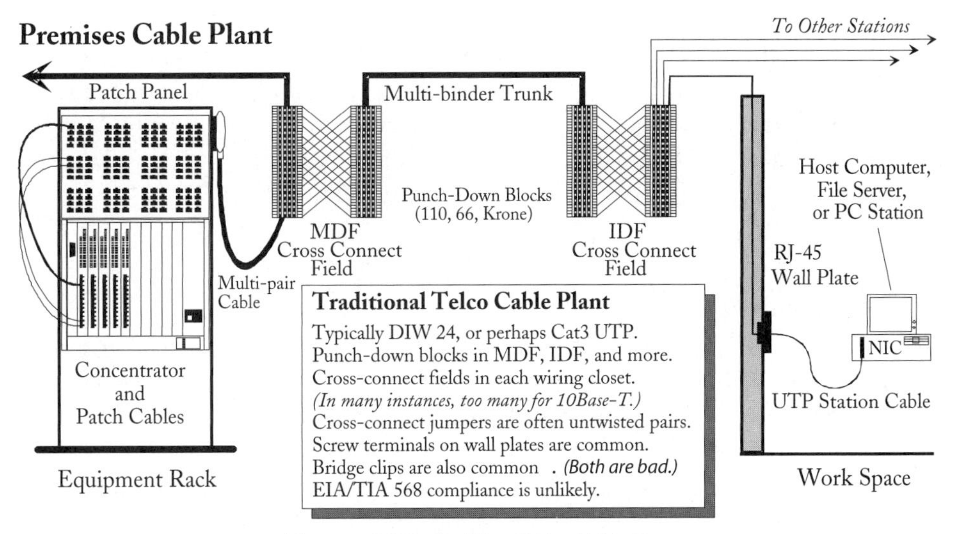

Figure 6.18 *Traditional Telco Cable Plant*

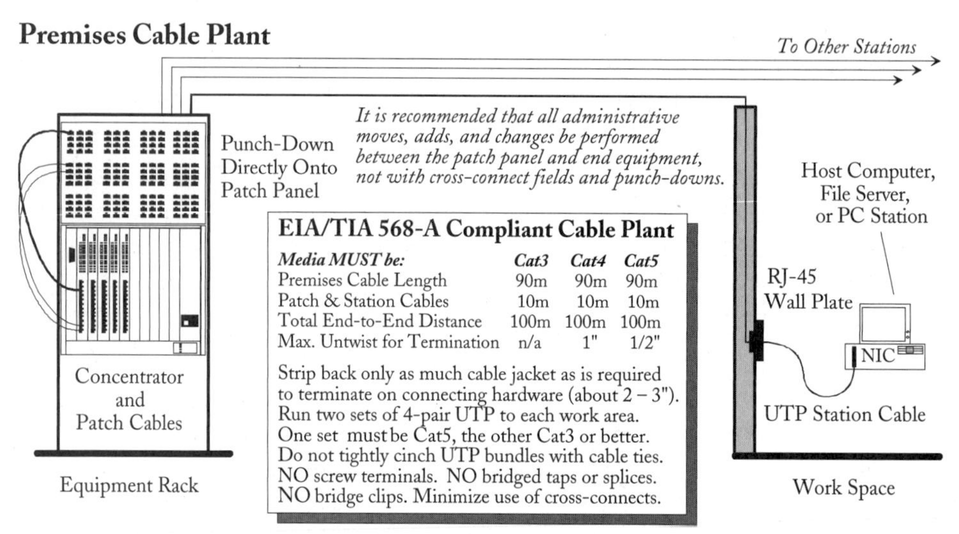

Figure 6.19 *TIA/EIA Standard Cable Plant*

The pin-out of the RJ-45 connectors at the patch panels and wall jacks must also be consistent. Most are color-coded to assist with installation, but there are several competing standards for RJ-45 pin-outs—two of which are provided by the same group of standards bodies! These two connector specifications are known as ANSI/TIA/EIA 568-A and 568-B. Mixing 568-A patch panels with 568-B wall jacks will result in un-pairing the conductors of pairs 2 and 3 (see Figure 6.20).

To prevent these problems from occurring, thoroughly document the cable plant design, adhere to the wiring standards, pick one standard to use throughout, use the correct patch panels and wall jacks, and avoid making "special" adaptor cables, or rewiring old, non-standard components and other such kludges. Do it right the first time, and don't cut corners—it costs less to do a thing correctly once than to do it over. The cable plant is the foundation of any network computing system; if it is not solid, everything built on top of it will be unstable and unreliable.

10BASE-F

PHYSICAL MEDIA

IEEE 10Base-F actually defines three separate 10 Mbps fiber standards: 10Base-FL, 10Base-FP, and 10Base-FB. The standards document was released as a supplement that states the objectives of the committee, proposes component definitions and new components, and identifies the media and connectors, distances supported, and intended configurations. The specification also proposed changes to much of the verbiage and the overall structure of the existing 802.3 standards documents, and offers more flexible methods of calculating round-trip collision delay and the worst-case path

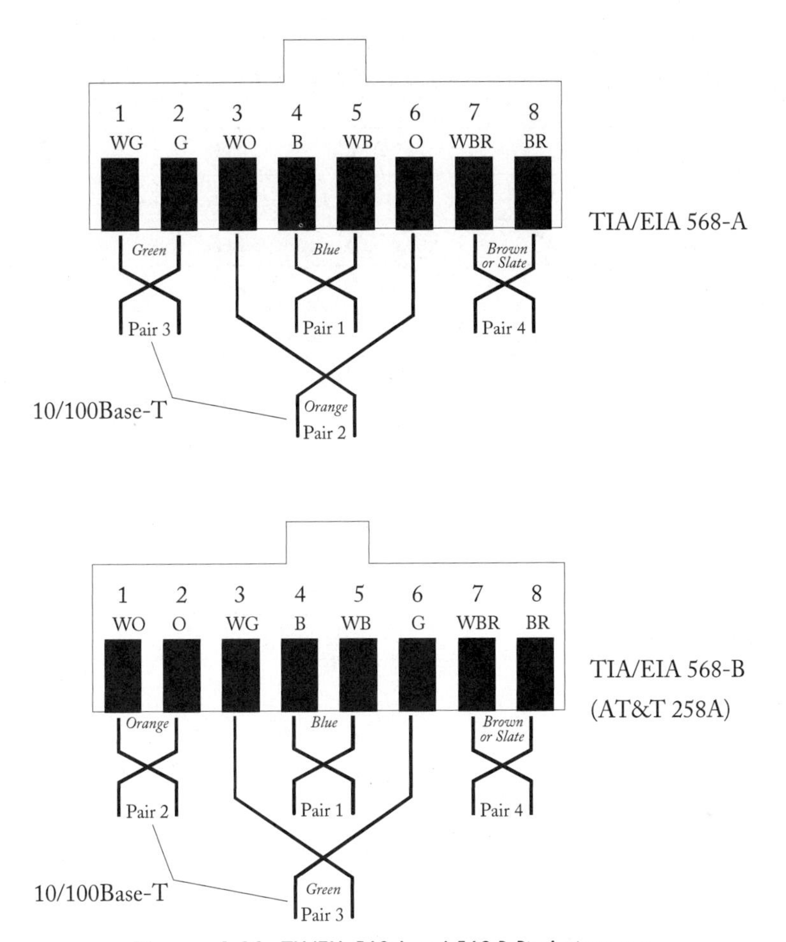

Figure 6.20 *TIA/EIA 568-A and 568-B Pin Assignments*

delay value to accommodate newer technologies (discussed in the chapter on repeaters). Table 6.5 gives specifications for optical fiber media used to support 802.3/Ethernet.

Fiber is specified by two measurements referenced in micrometers (μm). Multimode fiber is commonly available in two sizes: 50/125μm and 62.5/125μm. The first number defines the diameter of the fiber core, and the second number defines the diameter of the cladding material that surrounds the core. 10Base-F specifies duplex multimode fiber (see Figure 6.21).

Over the years fiber optic (FO) cabling has steadily gained popularity as an Ethernet medium. It has the greatest potential bandwidth, capable of supporting an organization's needs well into the future. Fiber cabling, components, and installation equipment have all become considerably less expensive in recent years. The media is extremely lightweight, smaller than any other media, supports a star-wired topology, provides the

TABLE 6.5 FIBER OPTIC MEDIA AND CONNECTOR SPECIFICATIONS AND CHARACTERISTICS

Connector	IEC BFOC/2.5
	AT&T Corp. ST equivalent
Approx. Connector Loss	0.4 dB (Mated to Fiber)
Maximum Insertion Loss	
10Base-FP	26 dB (@ 850 nm)
10Base-FB & FL	12.5 dB (@ 850 nm)

Cheaper, high-loss fiber may be used for short runs if other media specifications are met.

Fiber Size	62.5/125 µm
Alternate Media under Consideration	50/125 µm
Fiber Attenuation	< 3.75 dB/km (@ 850 nm)
Modal Bandwidth	> 160 MHz-km (@ 850 nm)
Minimum Velocity of Propagation	.67 c

Equivalent to a delay of no more than 5 µs/km.

Typical Bend Radius	38 mm (1.5")
Typical Weight per 1000'	12.3 lbs (Duplex Fiber)

greatest distances of any media, and is virtually impervious to electromagnetic and radio frequency interference or noise (EMI/RFI).

However, contrary to common wisdom, fiber is not as completely secure as some may have you believe. Rather than radiate electromagnetic signal energy like copper media, fiber radiates some of its infrared signal energy, and is therefore not impervious to eavesdropping. Certainly fiber makes it more difficult than does copper, but all it takes is more expensive eavesdropping hardware located in closer proximity to the media.

There are two basic types of optical fiber: single-mode and multimode. A mode is a ray of light that enters the core of the fiber at a particular angle (referred to as numerical aperture). Multimode fiber typically uses an LED as a light source, while single-mode fiber generally uses LASERs. LED technology typically produces a weaker signal across a broad range of modes, whereas LASER technology produces a much stronger signal in a single mode.

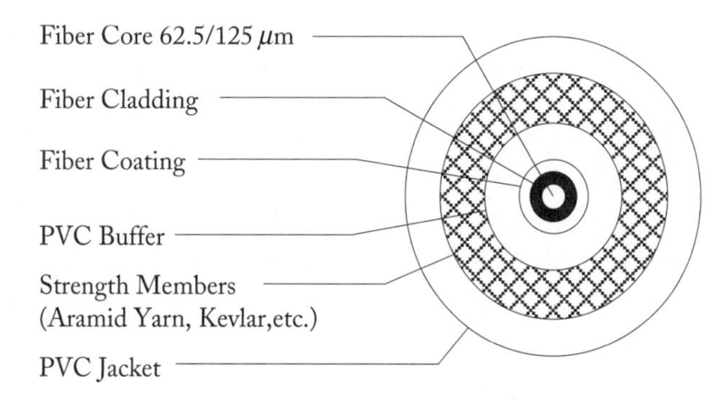

Optical Fiber Construction

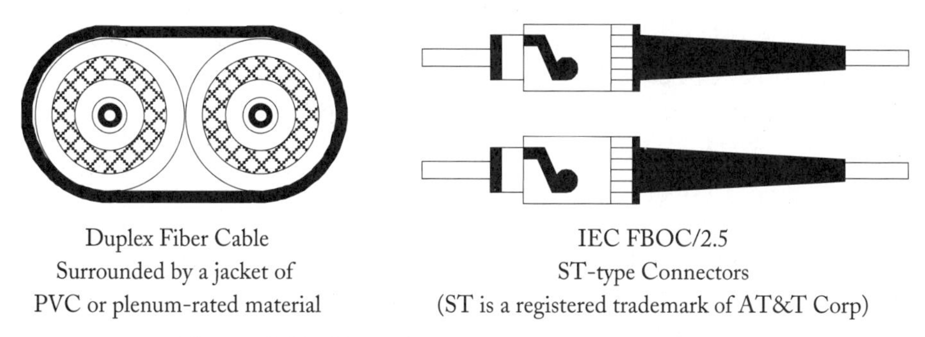

Duplex Fiber Cable
Surrounded by a jacket of
PVC or plenum-rated material

IEC FBOC/2.5
ST-type Connectors
(ST is a registered trademark of AT&T Corp)

Figure 6.21 *10Base-F Duplex Multimode Optical Fiber Cable*

The aperture of multi-mode fiber is large enough to accept multiple modes of light and allow them to propagate through the fiber. Because the modes of light enter the aperture of the fiber at different angles, each mode propagates through the fiber at a different rate, thereby causing them to arrive at the end of the fiber at different times. This phenomenon is called modal dispersion. Modal dispersion limits the bandwidth and distances that can be achieved using multi-mode fiber cabling. For this reason, multi-mode fiber is generally used for cable plants within a building or for other relatively small environments (<2–4 km).

INSTALLATION RULES

Optical fiber is always configured in a star or point-to-point. The 10Base-F standards specify only duplex multimode glass fiber, either 50/125 μm or 62.5/125 μm. Considerable latitude was provided for connector options, but most vendors use a miniature bayonet connector known as "stab and twist" (ST). For many years 62.5/125 μm was the most popular option, but it now appears that 50/125 μm may be surpassing it. There have also been technological advancements in fiber resulting in enhanced performance that has kept these two sizes of fiber media roughly on par with each other.

IEEE 10Base-F defines three new and different sets of optical fiber specifications and modifies an old one (see Figure 6.22).

10Base-FP defines a fiber-passive star and DTE transceiver. The passive hub design supports a network diameter of 1 km. Being passive hubs, they require no power. All light energy is supplied by the transceivers in each station/DTE. Each 10Base-FP fiber segment may extend as far as 500 meters from the passive hub to each station/DTE. Each passive hub may support up to 33 ports. Management is provided only on the DTE transceiver. This specification never achieved popularity.

10Base-FB defines a synchronous fiber backbone/repeater system and transceiver. The synchronous repeater implementation allows 10Base-FB to exceed the 2500-meter limitation of 10 Mbps Ethernet Collision Domains and four repeater hops. 10Base-FB can support Link Segments of 2 km. Due to the special synchronous signaling the 10Base-FB transceiver must be integrated into the repeaters. This specification saw very few applications.

10Base-FL defines a fiber Link Segment that supports both embedded transceivers as well as external transceivers. 10Base-FL supports Link Segments up to 2 km in length. Prior to the completion of the 10Base-F standard, many vendors implemented Ethernet over fiber using the older 802.3 fiber optic inter-repeater link specification (1988) which supported Link Segments limited to 1 km. Support for stations/DTEs was beyond the scope of the FOIRL specification. Although the intent of the FOIRL was to provide hub-to-hub connections only, many implementations supported DTE connections anyway. The new 10Base-FL standard supercedes the FOIRL and includes DTE support as well as hub-to-hub connections. 10Base-FL is also backward compatible with the FOIRL.

All fiber LAN technologies use duplex fiber—one strand of glass to transmit, the other to receive. As with 10Base-T, 10Base-FL supports a point-to-point or star topology, which makes it easy to troubleshoot. Fiber is a very flexible and robust medium, and has the greatest bandwidth potential, making it the most future-proof media available.

The biggest expense in fiber networking is not in the installation, repair, or maintenance of the fiber cable plant, but rather in the networking hardware. Optical fiber NICs, hubs, and switches are still expensive relative to their copper counterparts. In addition, due to a variety of factors the port density achieved with fiber hubs and switches is far below that of 10/100Base-T products. For example, where a 10/100Base-T hub or switch may provide twenty-four RJ-45 ports, a chassis of similar size may only be able to provide six or twelve fiber ports. Obviously, this can only exacerbate the cost disparity, and work is under way to improve the port density of fiber hub and switch products.

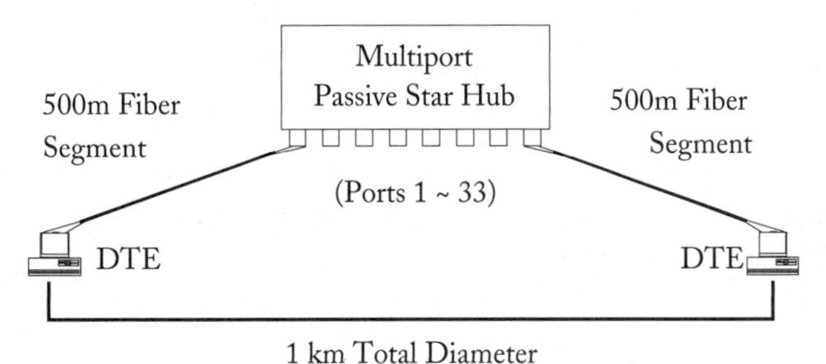

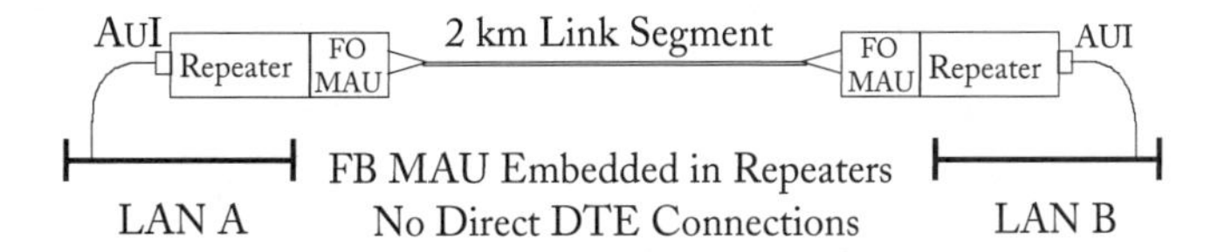

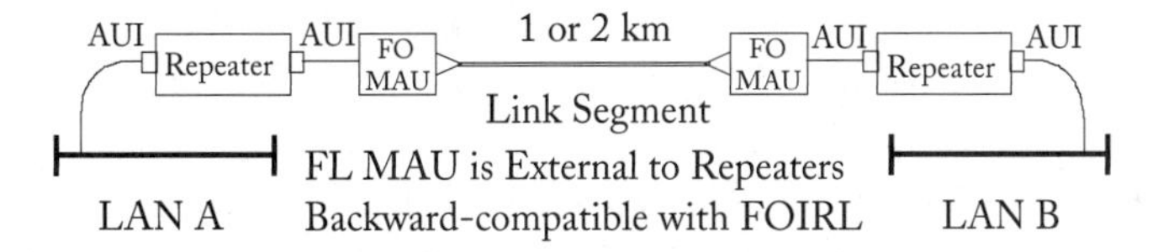

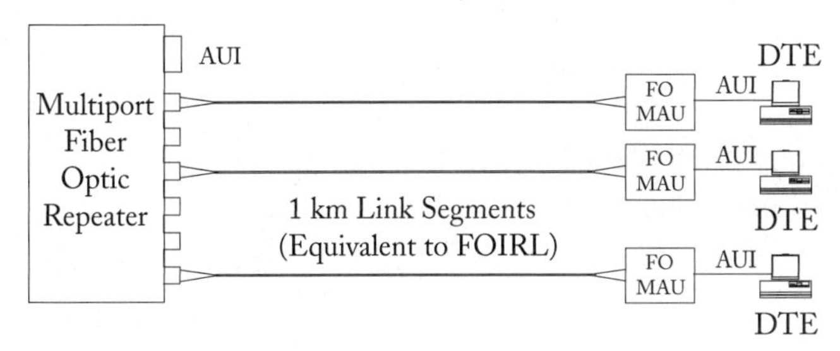

Figure 6.22 *10Base-F Installation Rules*

As with 10Base-T, 10Base-F structured cable plants can provide simple and economical moves, adds, and changes. Fiber repeater/hubs can provide superior maintenance, monitoring, and management capabilities. The light source employed in most of these FO cable systems is an infrared (IR) LED. IR LEDs consume little power and are small and relatively inexpensive, and the light generated is adequate for distances of at least 2 km.

Do NOT look into the end of a fiber cable! You may not know what is on the other end. If the light source is a LASER or a high-power LED, you may perform eye surgery on yourself! The light generated is not always in the visible part of the spectrum, and such intense light sources can seriously damage the retina of the eye. Wear protective eyewear and gloves when splicing fiber or installing fiber optic connectors. The glass can splinter as you work with it; the minute shards can penetrate the eyes and skin. With the splinters being so small and transparent, finding them can be almost impossible, and very painful.

CHAPTER SUMMARY

Since its creation in the late 1970s, Classic Ethernet has evolved to support new types of media and satisfy a diverse range of business requirements. The earliest forms of media were defined by the LAN standards themselves, whereas later LAN developments embraced the emerging cabling standards being independently developed. While these independent development efforts have yielded many benefits, they have also generated some confusion. In addition to added media flexibility, features and functionality have been built into the LAN technology itself to ensure reliable service while keeping the engineering elegantly simple. This synergy of technologies has resulted in the only LAN technology to have lasted more than twenty years and still be going strong.

REVIEW QUESTIONS

1. What is the difference between DIX ESPEC2 and IEEE 802.3?
2. List the primary and secondary functions of an Ethernet transceiver.
3. What type of media does 10Base-T support?
4. What is the maximum length of a 10Base-T Link Segment?
5. What are MDI and MDI-X and where are they used?
6. What is a Null Link Segment and where is it used?
7. Explain the difference between 568-A and 568-B connector pin-outs.

8. Define the following acronyms:

AUI	NIC
SQE	IFG
MDI	MDI-X
AWG	USOC
EMI	RFI

   Bonus acronyms: DTE, LED, LASER, FOIRL

9. What does link status prove?
10. How do repeaters propagate collisions?

Token Passing Rings

OBJECTIVES

After completing this chapter, you should be able to:

- Identify the key differences between token ring and CDDI.
- Explain the token passing network access protocol.
- Describe the functions of the token ring MSAU.
- Explain the purposes of SMP, AMP, and Neighbor Notification.
- Explain the functions of the phantom drive current.
- Describe the various STP cable types supported by token ring.
- Describe the differences between FDDI and CDDI.
- Define DAS, DAC, SAS.
- Identify the encoding schemes used by token ring, FDDI, and CDDI.
- Explain the purpose and process of Beaconing.

INTRODUCTION

Learn about the media used by alternative LAN technologies: token ring and FDDI/CDDI. This chapter will cover IEEE 802.5 token ring and ANSI X3T12 FDDI/CDDI, now referred to as legacy LANs.

IEEE 802.5 TOKEN RING

A Swedish engineer named Olof Soderblom is often credited with the invention of token-passing ring networks. In reality, Farmer and Newhall did some of the earliest work on token-passing rings during the 1970s at AT&T. Reportedly this early work was used as the basis for the Irvine Ring at the University of California, and later by Proteon in the development of its ProNET/10 product. As the story goes, IBM licensed the technology from Soderblom and used it to develop the IBM Token Ring Network at the company's labs in Zurich (the Zurich Ring). IBM continued this work in the United States.

In the early 1980s, the work on token ring technology was submitted to the IEEE 802 committee for consideration as a new LAN standard. Meanwhile, IBM developed the IBM Cabling System based on (150 Ohm) shielded twisted pair (STP) in a star-wired topology, and released the specifications in 1984. IBM customers started installing the cable plant even before the IEEE ratified token ring as a standard. By using *baluns,* the cable plant could be adapted to support existing IBM SNA terminal networks that traditionally use (93 Ohm) RG-62 coaxial cabling. Baluns are small cabling adapters that convert the balanced (150 Ohm) STP signal into an unbalanced (93 Ohm) coaxial signal. Some may correctly consider this approach to cabling a bit of a kludge, but this was just an interim solution to prepare the customer premises to use token ring as soon as it was available.

The first standard for 4 Mbps token ring over STP was released in late 1985 as ANSI/IEEE Standard 802.5. This LAN supports up to 250 stations and repeaters on one ring, and transmits using Differential Manchester Encoding (as does 10Base-T Ethernet). Token ring LANs use passive central wiring centers called Multi-Station Access Units (MSAU, or MAU). Being a passive device, the MSAU required no power, but it also provided no network management or monitoring capabilities (not even simple LED indicators). Despite the high price of token ring NICs, MSAUs, and cabling, many loyal IBM customers installed token ring LANs—even a few non-IBM customers were attracted to the promise of token ring and its deterministic access method.

Meanwhile, work proceeded on the development of a faster version of token ring. The IEEE ratified the 16 Mbps token ring in 1988, just three years after the first 4 Mbps version. Initially, both of these standards supported only STP cabling, but consumers demanded support for UTP instead. A few token ring vendors developed their own proprietary implementations, using a variety of UTP connectors and pin-out specifications, and eventually UTP support was added to the IEEE 802.5 standard (see Figure 7.1).

Eventually even IBM conceded to the demand for UTP support and developed several UTP-specific token ring products as well as media filters to connect STP components to UTP networks. Media filters are cabling adapters that provide impedance matching and frequency filtering, and convert the media connector from a (STP) DB-9 to a (UTP) RJ-45. Of course, media filters add cost to every token ring connection, and manufacturing separate STP and UTP products must have further reduced economies of scale. The result of all this was to keep token ring prices high and make token ring networking even more complicated (see Figure 7.2).

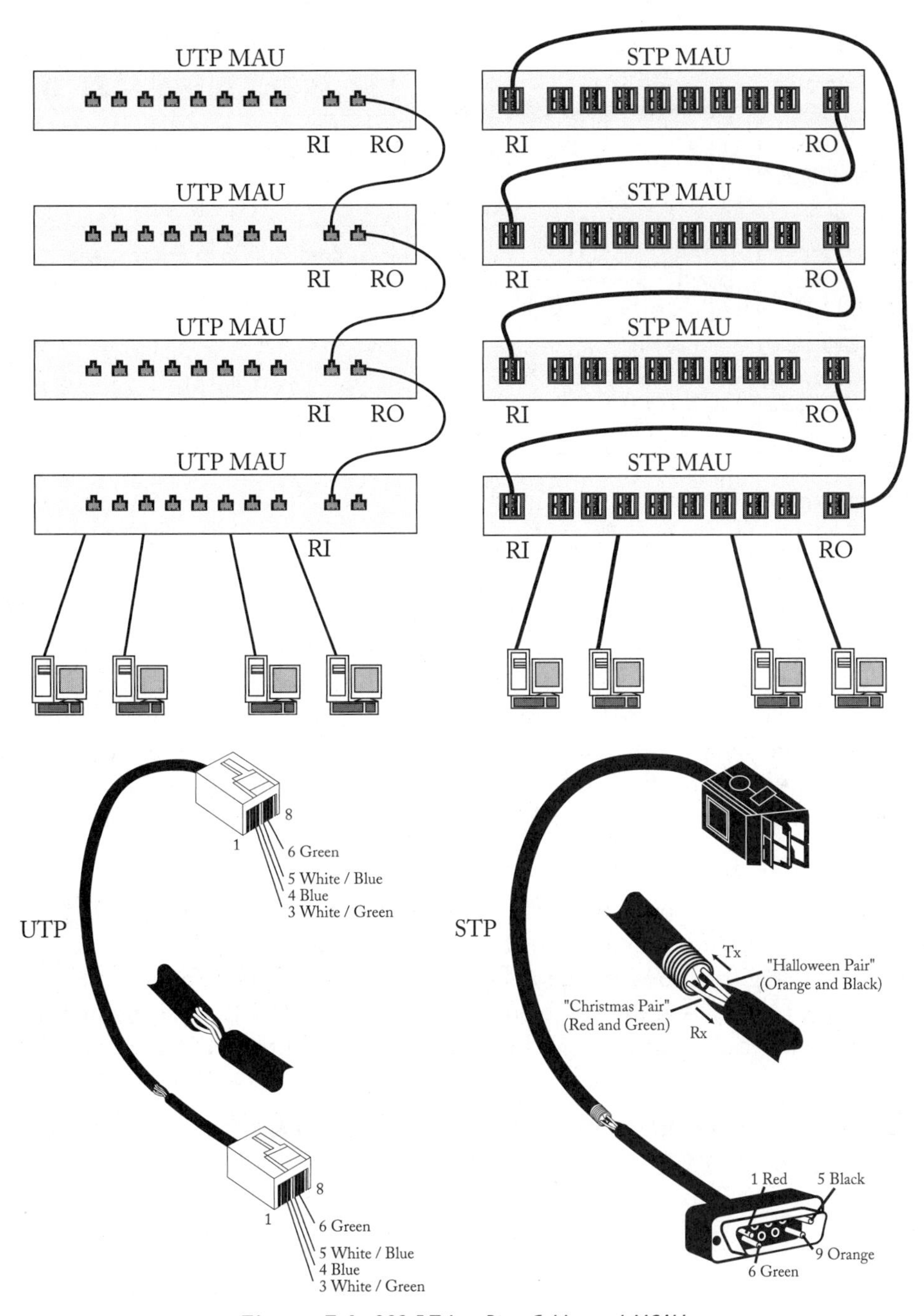

Figure 7.1 *802.5 Token Ring Cables and MSAUs*

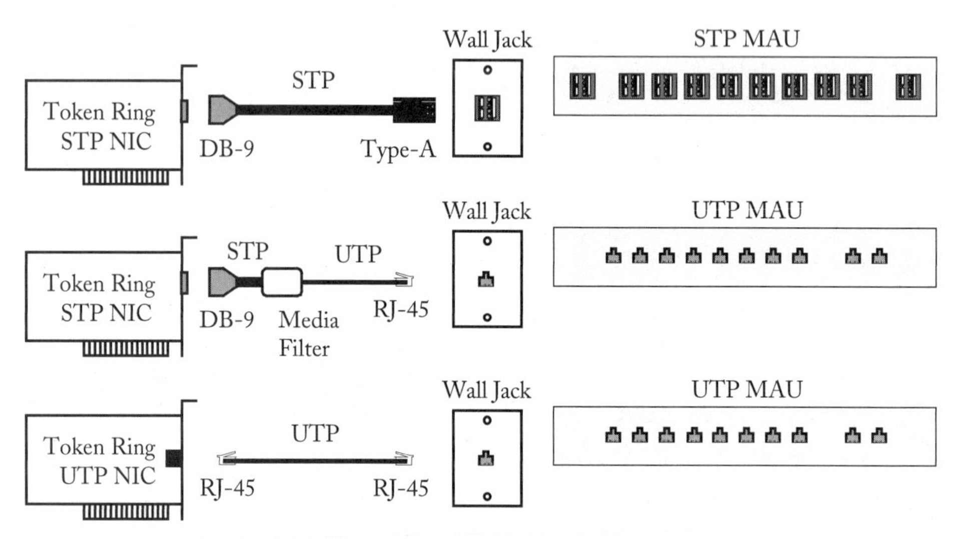

Figure 7.2 *Connections and Media Filters*

Consumers also insisted on a more intelligent and manageable alternative to the passive MSAU, which provides no management features at all. In response to consumer demand, several vendors developed active MSAUs. IBM responded with a pair of interdependent devices called the Control Access Unit (CAU) and Lobe Access Module (LAM). The management intelligence is contained in the CAU. Stations connect to lobe ports on the LAM. Each LAM supports twenty lobe ports, and each CAU can support a rack of four LAMs (see Figure 7.3).

PHYSICAL MEDIA

The original media supported by IEEE 802.5 token ring was IBM 150 Ohm STP. The IBM Cabling System defined the basic characteristics of its several different types of media, including (but not limited to) Type 1, Type 2, and Type 6. Support was later added for EIA/TIA standard Category 3, 4, and 5 UTP. IBM also produced its own UTP cable called Type 3, which is roughly equivalent to Category 3. The max-

TABLE 7.1 TOKEN RING COPPER MEDIA LOBE LENGTHS OPERATING AT 4 MBPS AND 16 MBPS USING PASSIVE AND ACTIVE MSAUS

Media	*Type 1, 2*	*Type 6*	*Type 9*	*Cat3 & 4*	*Cat5*
Active 4	300 m	30 m	200 m	200 m	250 m
Active 16	150 m	30 m	100 m	100 m	120 m
Passive 4	200 m	30 m	133 m	100 m	130 m
Passive 16	100 m	30 m	66 m	60 m	85 m

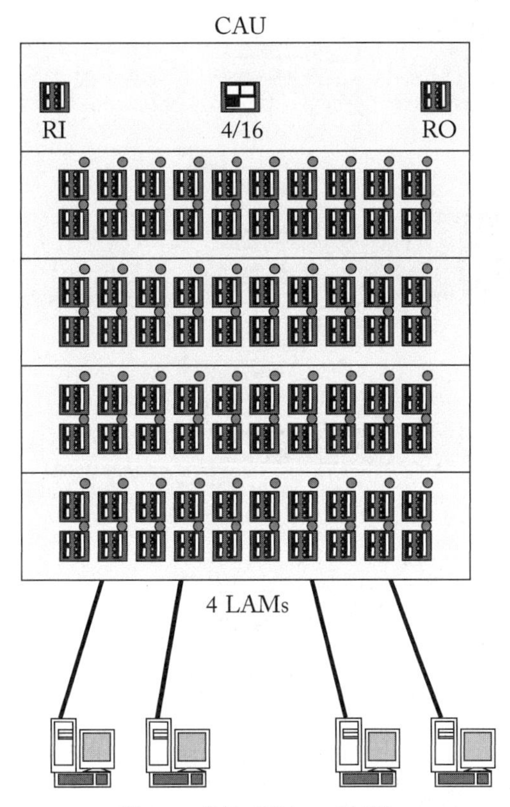

Figure 7.3 *CAUs and LAMs*

imum lobe length for either media depends on the type of media, the speed of the ring, and the use of active or passive MSAUs. Table 7.1 shows the maximum lobe lengths for various types of copper token ring media when operating at 4 Mbps and 16 Mbps using either passive or active MSAUs (see Figure 7.4).

STP Specifications

Table 7.2 gives specifications for shielded twisted pair copper media used to support token ring.

Table 7.3 compares the original family of IBM cable types and some of their attributes.

The following paragraphs describe each type of copper media in the IBM family of twisted pair cables.

Type 1

Two STPs consisting of solid 22 AWG copper conductors. Each pair is individually wrapped in a foil shield, and a heavy braided metal shield surrounds both pairs. Typically used for the longest premises cable runs. Intended for data only.

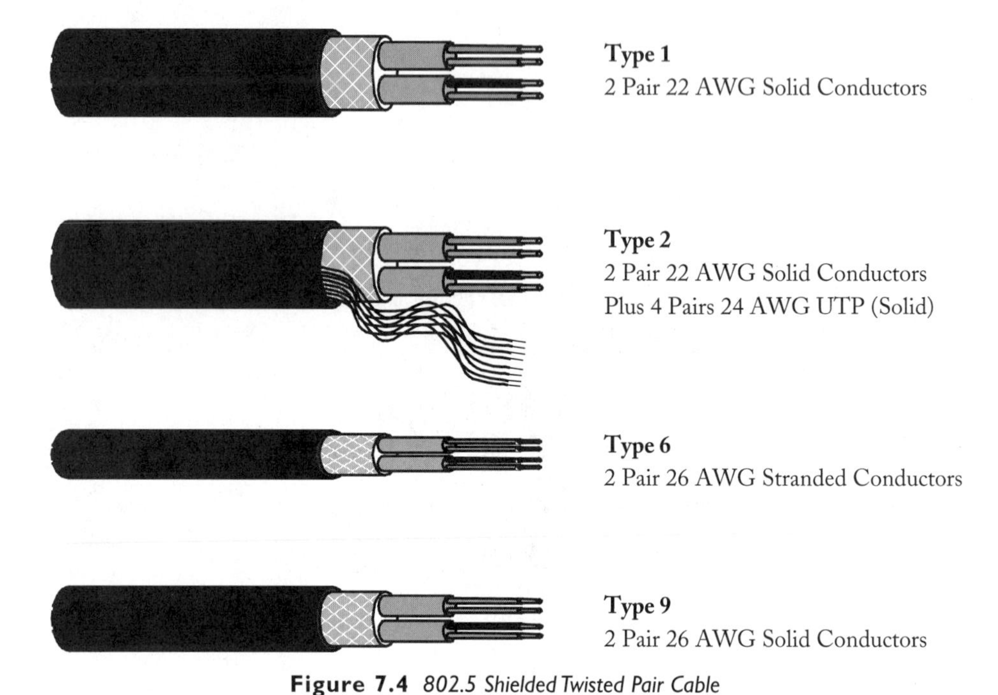

Figure 7.4 *802.5 Shielded Twisted Pair Cable*

TABLE 7.2 SHIELDED TWISTED PAIR MEDIA SPECIFICATIONS AND CHARACTERISTICS

Differential Characteristic Impedance [7.3]	150 ±15 Ohms @ 2–20 MHz	
Channel Attenuation at 4 and 16 Mbps [7.5.2]:	4 Mbps	16 Mbps
Square Root Attenuation (SQA)	22 dB	19 dB
Flat Attenuation	15 dB	15 dB
Attenuation (Type 1, 2)	≤22 dB/km	≤45 dB/km
Attenuation (Type 6, 9)	≤33 dB/km	≤66 dB/km
Connector Insertion Loss (150 Ohms) [7.9]	< 0.1 dB @ 100 kHz to 16 MHz	
Crosstalk Rejection	> 62 dB @ 100 kHz to 4 MHz	
Crosstalk Rejection	> 50 db @ 100 kHz to 16 MHz	

Type 2

A hybrid cable similar to Type 1 data cable, but with four additional UTPs consisting of solid 24 AWG copper conductors. The additional pairs are outside of the

TABLE 7.3 THE ORIGINAL FAMILY OF IBM CABLE TYPES AND THEIR ATTRIBUTES

Cable Type	*1*	*2*	*3*	*6*	*8*	*9*
Gauge	22	22/26	24	26	26	26
Shielded	X	X		X	X	X
Unshielded		X	X			
Plenum	X	X	X			X
Non-plenum	X	X	X	X	X	
Outdoor	X		X	X		
Voice and Data		X	X			
Data only	X		X	X	X	X

shielding and are intended to support voice communication, but they will support 10Base-T. Type 2 was designed to deliver voice and data using one cable run.

Type 3
Usually four UTPs consisting of solid 24 AWG copper conductors intended for low-speed data or voice communications, such as telephone lines. Also used for premises cabling and will also support 10Base-T.

Type 6
Two STPs consisting of stranded 26 AWG copper conductors instead of solid, intended for use as patch or station cables. Type 6 is a smaller-gauge STP that is more flexible, making it easier to handle. However, due to higher attenuation, Type 6 cable limited to two-thirds the cable distance of Type 1.

Type 8
Two flat STPs of 25 AWG stranded wire for under-carpet installation. Type 8 STP is supposed to have the same characteristics as Type 9 cabling, but it usually does not.

Type 9
Similar to Type 1 and Type 8, but consists of solid 26 AWG copper conductors. Again, due to higher attenuation, Type 9 cable is limited to two-thirds the cable distance of Type 1.

UTP Specifications

Table 7.4 gives specifications for unshielded twisted pair copper media used to support token ring.

TABLE 7.4 UNSHIELDED TWISTED PAIR MEDIA SPECIFICATIONS AND CHARACTERISTICS

Characteristic Impedance	*85–111 Ohms*
Attenuation—Category 3	≤56 dB/km @ 4 MHz, ≤ 131 dB/km @ 16 MHz
Attenuation—Category 4	≤42 dB/km @ 4 MHz, ≤ 88 dB/km @ 16 MHz
Attenuation—Category 5	≤42 dB/km @ 4 MHz, ≤ 82 dB/km @ 16 MHz
Near End Crosstalk—Category 3	23 dB/1000 feet
Near End Crosstalk—Category 4	36 dB/1000 feet
Near End Crosstalk—Category 5	44 dB/1000 feet

CABLES AND CONNECTORS

STP token ring uses special hermaphroditic plugs and jacks called the Medium Interface Connector (MIC) by the IEEE standards, or the Universal Data Connector (UDC) in IBM parlance. Unlike all other network connectors, the MIC or UDC does not rely on a male plug and a female jack; the exact same connector is used as both plug and jack. Any MIC connector can be rotated 180 degrees to plug directly into another MIC. In addition, the MIC provides a built-in loopback function. When a MIC cable is disconnected from an MSAU, patch panel, or wall plate, a bar inside the MIC closes a circuit between the transmit pair onto the receive pair. The purpose of this function is to facilitate testing and troubleshooting. The MIC also provides a locking mechanism that does away with the need for screws. Unfortunately, over time the locking tabs can begin to wear and the weight of the cable pulls the MIC from its port (usually at the most inopportune time). While the MIC is a very creative piece of engineering, it never became economical. In addition, because personal computer adapter slots are so much narrower than the MIC, token ring NICs could not support the MIC and were instead fitted with the more common (female) DB-9 jack. To interface the NIC to the token ring cable plant, station cables had to be assembled with a MIC on one end and a (male) DB-9 plug on the other. The MIC is defined in sections [7.9], [7.9.1], and [7.9.2] of the 802.5 standard (see Figure 7.5).

The original token ring MSAU is a passive device that typically supports eight lobe ports and a pair of ring-in (RI) and ring-out (RI) ports. Stations/DTEs connect to the eight lobe ports, but merely attaching the station cable to a lobe port on the MSAU does not provide access to the ring. Station insertion into the ring is controlled by the NIC in the station/DTE. Each of the lobe ports on an MSAU provides the interface to a Trunk Coupling Unit (TCU). The TCU provides the mechanism responsible for effecting the bypass or insertion of the station into the ring.

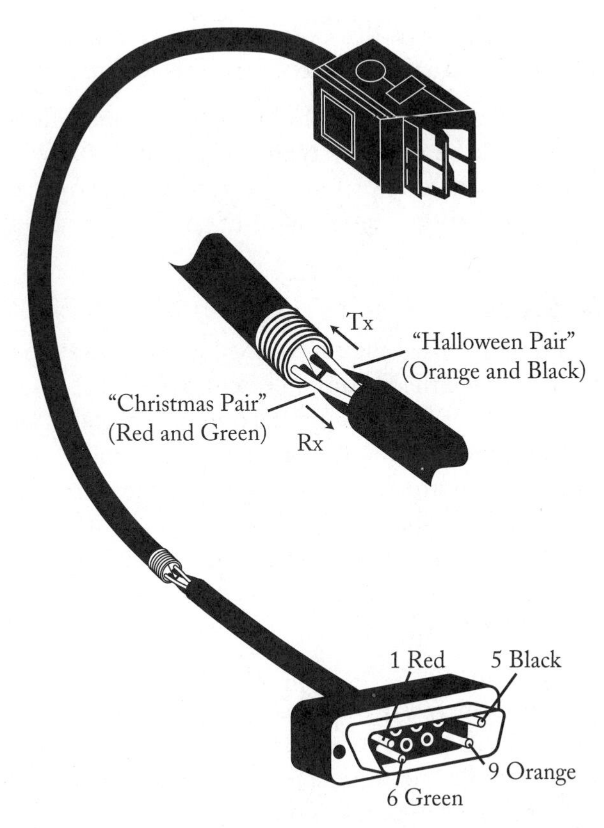

Figure 7.5 *802.5 STP Medium Interface Connector*

Each TCU is a bypass relay that must be energized by the station attached to that port. Rather than use additional copper pairs in the station cable to deliver the power needed to energize the bypass relay, a phantom drive circuit technique impresses a DC voltage on the data conductors. The DC voltage is transparent to transmitted data, hence the term *phantom*. If for any reason the phantom voltage is terminated (the station is powered down or a cable is cut), the TCU on that MSAU port will close its bypass relay to circumvent the lobe servicing the faulty station, and thereby maintain the integrity of the ring.

When first installing an MSAU, each lobe port must be reset to its correct default position. As counter-intuitive as that may sound, it is possible during shipment to jostle the MSAU such that the bypass relays may be flipped into the wrong position. If this happens, it will not be possible to complete the ring circuit (unless perhaps all lobe ports are in use). To overcome this problem, a Token Ring setup tool is used on each lobe port to reset each to the correct default position. The Token Ring setup tool is merely a MIC on a stick with an LED and a 9-volt battery in the handle. The tool is inserted into each successive lobe port. When the tool is inserted, the LED will illuminate and slowly fade until a *click* is heard. A brief flash of the LED usually accompanies the click. That click is the sound of the bypass relay resetting to the correct default position. If this sounds archaic to you, it is because you are correct.

The ring-in and ring-out ports are used to interconnect one MSAU to another. It is recommended that there be no more than ten (10) RI/RO devices on the same token ring. The RI/RO ports do not interface with a TCU and therefore provide no inherent bypass capability. Stations/DTEs may *never* be attached to the RI/RO ports.

INSTALLATION RULES

Before UTP was supported, designing token ring networks to support STP required a detailed formula to calculate the maximum extent of each LAN's topology. Instead of defining a fixed number of stations and maximum lobe cable length, token ring topology extent is determined by the type of media used, length and number of lobe cables, data rate, maximum frame size (which is dependent on data rate), and number of MSAUs, inter-MSAU cables, and the limitations of the token rotation timer. As more stations or MSAUs are added to the ring, the maximum permissible lobe length is reduced. Conventional token ring cable plant design requires a clear vision of future growth expectations.

One of the major problems with this approach is the fact that networks are not static; rather, they are extremely dynamic systems. Initial design calculations quickly become obsolete as soon as most networks are put into service. Fortunately, the arrival of UTP vastly simplified network engineering. Instead of relying on a complex formula to design dynamic networks, token ring networks using UTP may be designed based on the EIA/TIA 568A standards for structured cabling systems. Simplified rules for STP are also provided.

As mentioned above, the maximum lobe length for STP depends on the type of STP media, the speed of the ring, and the use of active or passive MSAUs. Type 1 and Type 2 STP lobe cables can support lengths up to 300 meters (984 feet) at 4 Mbps, or 150 meters (492 feet) at 16 Mbps. Using passive MSAUs, those lobe lengths are reduced to 200 meters (656 feet) and 100 meters (328 feet), respectively.

Type 9 STP lobe cables can support lengths up to 200 meters (656 feet) at 4 Mbps, or 100 meters (328 feet) at 16 Mbps. Using passive MSAUs, those lobe lengths are reduced to 133 meters (436 feet) and 66 meters (216 feet), respectively. Type 6 STP is always limited to 30 meters (98 ft) and should only be used as patch or station cables.

IBM Type 1 is the only STP cable recommended for RI/RO trunk cables. Type 1 STP trunk cables can support lengths up to 770 meters (2525 feet) at 4 Mbps, or 346 meters (1134 feet) at 16 Mbps.

Each of the above-mentioned STP cable types enjoys a newer enhanced version identified by the addition of the letter "A" as a suffix to the media type. STP types 1A, 2A, 6A, and 9A are each rated up to 300 MHz rather than the original 16 MHz. This specification provides added spectral headroom (greater bandwidth) and is required to support 100 Mbps high-speed token ring.

The maximum attenuation for any token ring UTP cable link is dependent on the speed of the ring. 16 Mbps rings use 16 MHz of bandwidth and therefore must meet more stringent cabling criteria than 4 Mbps rings that use 4 MHz of bandwidth. Attenuation calculations must include all cabling components in the cable path, including premises cabling, patch and station cables as well as the patch panels themselves, punch-down blocks, and wall jacks.

As with STP, the maximum UTP lobe length also depends on several criteria. Category 3 or Category 4 lobe cables can support lengths up to 200 meters (656 feet) at 4 Mbps, or 100 meters (328 feet) at 16 Mbps. Using passive MSAUs those lobe lengths are reduced to 100 meters (328 feet) and 60 meters (196 feet), respectively.

Category 5 UTP cabling, which has superior electrical characteristics, is capable of supporting longer link lengths than Category 3 or 4. Category 5 lobe cables can support lengths up to 250 meters (820 feet) at 4 Mbps, or 120 meters (393 feet) at 16 Mbps. Using passive MSAUs reduces the maximum lobe length to 130 meters (426 feet) at 4 Mbps, or 85 meters (278 feet) at 16 Mbps. UTP cable is not recommended for RI/RO (trunk) cables.

One last annoying detail: most of the token ring cabling rules violate the EIA/TIA 568-A cabling standards. The EIA/TIA defines the electrical characteristic for the categories of UTP media, as well as the standards for design and installation. According to the EIA/TIA standards, all categories of UTP are limited to a 100-meter channel. Furthermore, Category 4 has been discontinued, and Category 5 has been superceded by Category 5E. This topic will be covered in detail in a subsequent chapter.

IEEE 802.5j added support for duplex multi-mode and single-mode optical fiber. Single-mode fiber supports greater distances than multi-mode fiber, both are available in a variety of sizes, and each size has its own attenuation characteristics. The following tables give the attenuation specifications for each type of optical fiber.

Multi-mode fiber attenuation (Based on a wavelength of 850 nm):

-13 dB or less for 50/125 μm optical fiber
-16.0 dB or less for 62.5/125 μm optical fiber
-19.0 dB or less for 100/140 μm optical fiber

Single-mode fiber optic links may be 8.3/125 μm or 12/140 μm fiber. The maximum length of a single-mode fiber optic station cable is 10 km (32,800 ft). The maximum length of a single-mode fiber optic trunk cable is identical to the maximum allotment for station connections using the same media.

Single-mode fiber attenuation (Based on a wavelength of 1,300 nm):

-15.1 dB or less for 8.3/125 μm fiber
-15.1 dB or less for 12/140 μm fiber

ANSI X3T12 FDDI AND CDDI

Originally conceived in the late 1970s as a high-speed point-to-point interface between mainframes and peripherals, the Fiber Distributed Data Interface (FDDI) was to become the dominant high-performance LAN of the 1990s. Olof Soderblom held the patents for just about everything token ring, including FDDI. Since Olaf received royalties (about 3%) for every token ring thing ever sold, several token ring and FDDI vendors brought the matter of patent rights before the courts. The courts initially ruled in favor of the plaintiffs, but those decisions were eventually overturned on appeal. Many ignored Olaf's patents until they finally expired in 1998.

The first FDDI standard was released in 1988 as ANSI X3T9.5, and was renamed to ANSI X3T12 in 1995. The complete FDDI specifications actually consist of a collection of ANSI/ISO standards that define the various components of the protocol. The four key components of the FDDI standards include: Media Access Control (MAC), Physical layer Protocol (PHY), Physical layer Medium Dependent (PMD) interface, and the Station Management (SMT) protocol (see Figure 7.6). Underlying these primary protocol components are the unique specifications for each of the different supported media, including Twisted Pair PMD (TP-PMD), single-mode Fiber PMD (SMF-PMD), and multi-mode fiber PMD (MMF-PMD).

MAC handles LAN addressing, scheduling, and frame forwarding, and is defined by ANSI X3.139-1987/ISO 9314-2:1989.

PHY handles data encoding/decoding, NRZI modulation, and clock synchronization, and is defined by ANSI X3.148-1988/ISO 9314-1:1989.

PMD handles the analog baseband transmission over both fiber and copper media, and is defined by ANSI X3.166-1990/ISO 9314-3:1990.

SMT handles ring management including neighbor identification, fault detection, and reconfiguration, and is defined by X3.229-1994/ISO 9314-6:199x. SMT is a low-level protocol that provides link-level management for FDDI by addressing the management of functions provided by MAC, PHY, and PMD. It performs functions such as ring recovery, frame level management, link control, etc. All FDDI stations support SMT.

Other related standards include:

Single-Mode Fiber PMD (SMF-PMD), ANSI X3.184-1993, ISO 9314-4:199x

Low-Cost Fiber PMD (LCF-PMD), X3.237-1995, ISO 9314-9:199x

Twisted Pair PMD (TP-PMD), X3.263.1995, ISO 9314-10:199x

Physical layer Repeater (PHY-REP), X3.278-199x

ISO OSI	ANSI X3T12 FDDI			
LLC	IEEE 802.2 Logical Link Control			
MAC	FDDI Media Access Control (MAC)			FDDI Station Management (SMT)
Physical	FDDI Physical Layer Protocol (PHY)			
	FDDI Physical Layer Media Dependent (PMD)			
	Twisted Pair PMD	Single-mode Fiber PMD	Multimode Fiber PMD	

Figure 7.6 *ANSI X3T12 FDDI/CDDI Protocols*

FDDI development was based on IEEE 802.5 token ring. Whereas token ring uses a priority/reservation token access method, FDDI uses a timed token protocol. Therefore, there are several differences in frame formats and how station traffic queuing is handled, and significant differences in ring management. Like token ring, an FDDI token is a special 3-octet frame (following the FDDI preamble). FDDI stations must wait for a free token, grab the token, transmit one or more frames, and release the token. Timers in the MAC protocol hardware determine the number of frames that can be transmitted before the token must be released.

FDDI LANs support up to 500 stations on one ring, and transmits using 4B5B encoding. FDDI is typically configured in a star-wired topology based on active central wiring centers referred to as concentrators. The maximum length for a dual FDDI ring is 100 km (62 miles), or a total of 200 km (124 miles) for both rings, with a maximum of 2 km between stations (see Figure 7.7).

The most common applications for FDDI have been to provide high-speed server-farm interconnections and as fault-tolerant campus backbones interconnecting numerous slower 10 Mbps Ethernet and 4/16 Mbps token ring LANs. The primary application for CDDI is to connect many SAS workstations to concentrators that are then interconnected via DAS to an FDDI backbone.

The grand plans of FDDI's proponents never achieved fruition. There were proposals to extend the FDDI data rate to 600 Mbps, and to develop an isochronous data-transmission capability to support voice and video as well as data. The new

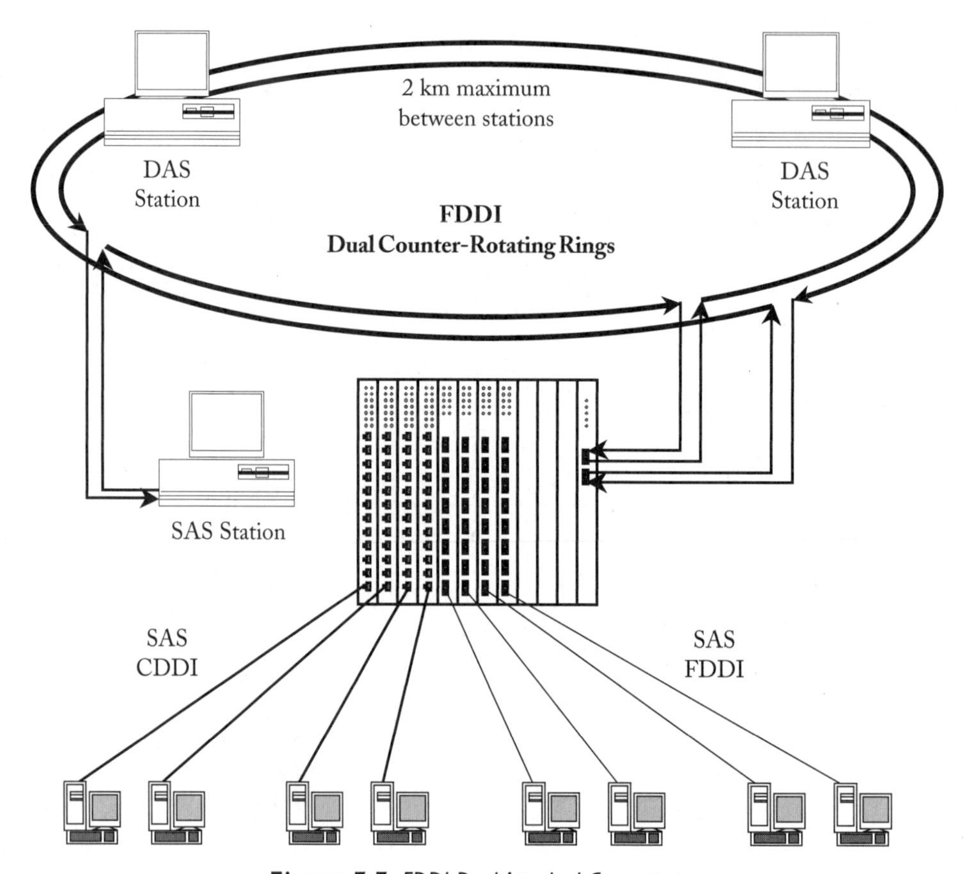

Figure 7.7 *FDDI Dual Attached Concentrator*

implementation was to be called FDDI-II, and was to be backward compatible with the original FDDI standard; however, FDDI-II never saw the light of day.

PHYSICAL MEDIA

FDDI supports duplex multi-mode and single-mode optical fiber using a variety of different connectors. Multi-mode fiber optic Link Segments may be 50/125 μm, 62.5 μm, or 100/140 μm. Single-mode fiber optic Link Segments may be 8.7/125 μm. Fiber optic budget is the combination of optical loss of the fiber optic cable, inline splices, and other fiber optic connectors (see Figure 7.8).

As with token ring and other technologies, single-mode fiber supports greater distances than multi-mode fiber, both are available in a variety of sizes, and each size has its own attenuation characteristics. Tables 7.5 and 7.6 give the attenuation and light loss budget specifications for multi-mode fiber. Table 7.7 gives the attenuation specifications for single-mode fiber.

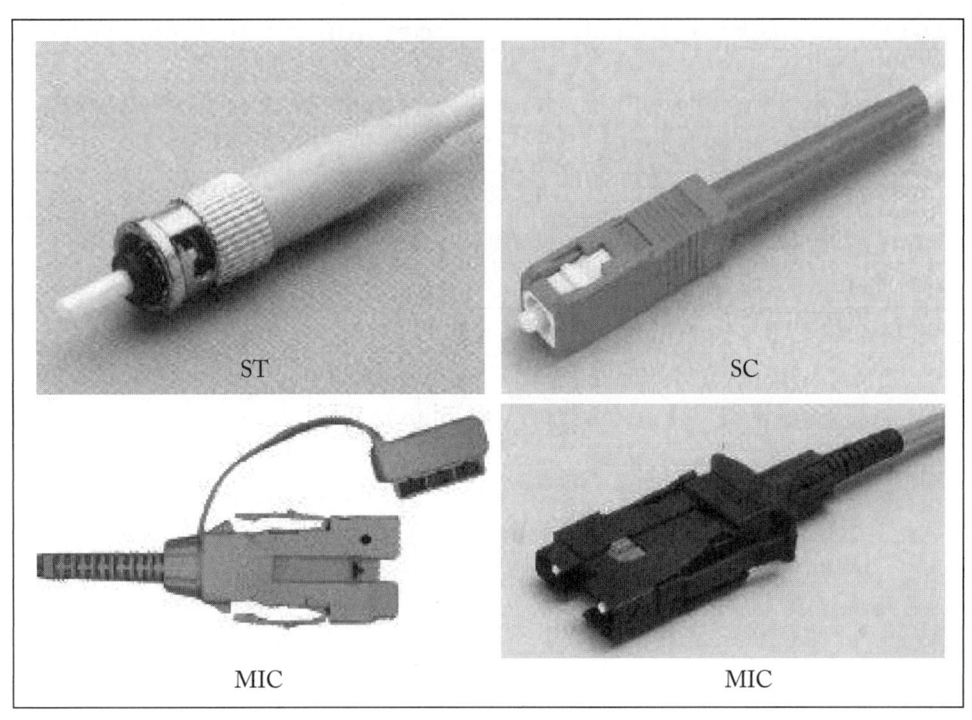

Figure 7.8 *FDDI Duplex Multi-Mode and Single-Mode Optical Fiber Cable*

TABLE 7.5 MULTI-MODE FIBER SPECIFICATIONS

Properties	***Specifications***	
Multi-mode fiber attenuation	13.0 dB or less for 50/125 μm optical fiber	
(Based on a wavelength of 850nm)	16.0 dB or less for 62.5/125 μm optical fiber	
	19.0 dB or less for 100/140 μm optical fiber	
Maximum cable segment length	Multi-mode	2 km (6562 feet)
	Single-mode cat-I	14 km (8.4 miles)
	Single-mode cat-II	58 km (34.8 miles)

Over the years the Copper Distributed Data Interface (CDDI) has supported STP and a variety of UTP copper cabling, including IBM Type 1 STP (up to 150 meters), DIW24 UTP (up to 50 meters), and high-performance data-grade UTP such as Category 5 up to 100 meters (328 feet). CDDI employs the TP-PMD interface, which defines 4B5B encoding with MLT 3.

TABLE 7.6 MULTI-MODE FIBER LIGHT LOSS BUDGET SPECIFICATIONS

Parameter	*Typical*	*Worst Case*	*Worst Case Budget*	*Typical Budget*
Value				
Receive Sensitivity	–30.5 dBm	–28.0 dBm	---	---
Peak Input Power	–7.6 dBm	–8.2 dBm	---	---
Transmit Power				
50/125 μm	13.0 dBm	–15.0 dBm	13.0 dB	17.5 dB
62.5/125 and 50/125	10.0 dBm	–12.0 dBm	16.0 dB	20.5 dB
100/140 and 50/125	7.0 dBm	–9.0 dBm	19.0 dB	23.5 dB
Error Rate:	Better than 1010 bit error rate			

TABLE 7.7 SINGLE-MODE FIBER SPECIFICATIONS

TCH: Properties	*Specifications*
Single-Mode Fiber Attenuation (Based on a wavelength of 1305 nm)	–0.5 dB or less for 8.7/125 μm optical fiber *The fiber core diameter may be 8–12 μm*
Maximum Cable Length	40 km (24.85 miles), if fiber optic budgets are met
Spectral Width	15 nanometers maximum
Optical Risetime	2 nanoseconds
Optical Falltime	2 nanoseconds
Transmitter Output	–14 dBm maximum, –20 dBm minimum
Receiver Input	–14 dBm maximum, –31 dBm minimum

CABLES AND CONNECTORS

FDDI uses special duplex fiber connectors that are keyed to define their media type and purpose. FDDI keying types distinguish between multi-mode fiber and single-mode fiber, for dual attachment station (DAS), single-attachment station (SAS), and dual-attachment concentrator (DAC) (see Figure 7.9).

CDDI uses the typical 8-position modular jacks commonly referred to as RJ-45. Two copper pairs are required, one pair to transmit and the other to receive. Unlike Ethernet or token ring, CDDI uses the two outermost pairs of the RJ-45 connector, EIA 568-B pairs 2 and 4 (EIA 568-A pairs 3 and 4) (see Figure 7.10).

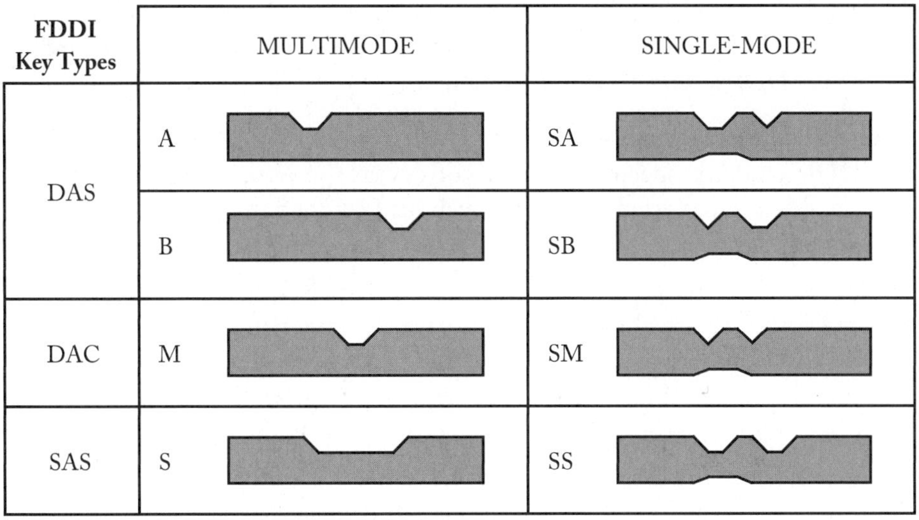

FDDI Key Types	MULTIMODE	SINGLE-MODE
DAS	A	SA
DAS	B	SB
DAC	M	SM
SAS	S	SS

Figure 7.9 *FDDI Connector Key Types*

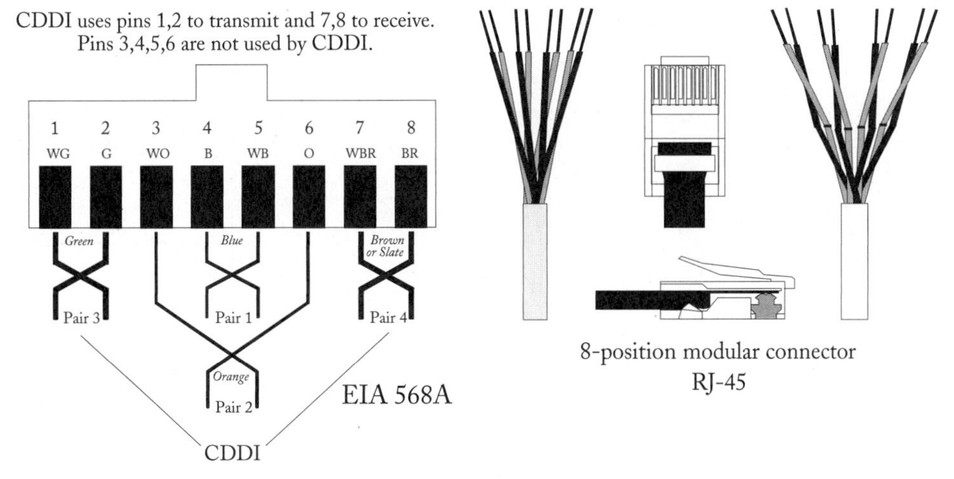

Figure 7.10 *CDDI Cables and Connectors*

INSTALLATION RULES

The installation details for FDDI and CDDI are considerably more complicated than any flavor of Ethernet or token ring. Much of the added complexity derives from the redundant dual ring topology. In addition to the inherent fault tolerance provided by the dual rings, FDDI supports more than one way to attach a station. To leverage the fault tolerance requires that stations be attached to both rings, which can be expensive. To save money, less critical stations may be attached to just one of the rings. To ensure the highest reliability requires optical bypass switches and/or dual homing, both of which add even more cost to each connection. The following sections will address the various FDDI station attachment methods.

Station Attachment Types

The FDDI standards specify four types of device connections: single-attachment station, dual-attachment station, single-attachment concentrator, and dual-attachment concentrator.

SAS devices attach only to the primary ring via an FDDI concentrator. Devices connected with SAS attachments will not have any effect on the FDDI ring if they are disconnected, powered off, or suffer a failure. FDDI concentrators serve as wiring hubs for multiple FDDI attached stations and will be covered in more detail.

DAS devices have two interfaces, designated A and B. These interfaces connect the DAS device to both the primary and secondary FDDI rings. Devices using DAS connections will affect the FDDI rings if they are disconnected, powered off, or suffer a failure. Figure 7.11 shows FDDI DAS A and B interfaces with attachments to the primary and secondary rings.

An FDDI concentrator or DAC attaches directly to both the primary and secondary FDDI rings. Disconnecting, powering off, or suffering a failure by any SAS device does

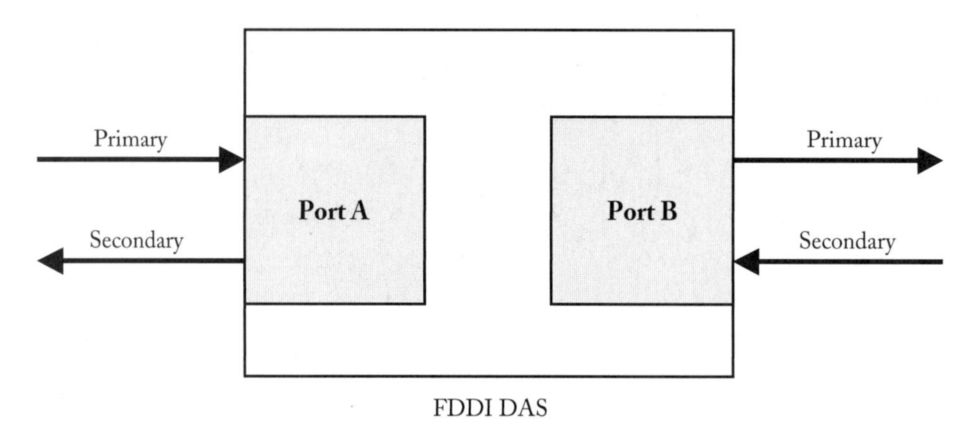

Figure 7.11 *Dual Attachment Station Interfaces*

not bring down the entire ring. The use of FDDI concentrators is particularly useful when connecting devices such as PCs to the ring, or connecting similar devices that are frequently powered on and off. Figure 7.12 shows the FDDI ring attachments of SAS and DAS stations, and DAC concentrators.

Dual Ring Fault Tolerance

FDDI provides a number of unique fault-tolerant features, including the dual ring topology, support for optical bypass switches, and dual-homing capability. The primary fault-tolerant feature of FDDI is its support for dual rings. If a DAS attached device is disconnected, powered off, suffers a failure, or is stolen (which has happened!), the primary ring is automatically wrapped onto the secondary ring. When the ring is wrapped, the dual-ring topology becomes a single-ring topology, and thereby loses its fault tolerance. During the wrap condition, data transmissions on the FDDI LAN continue uninterrupted with no impact on performance. Figures 7.13 and 7.14 illustrate a wrapped FDDI ring.

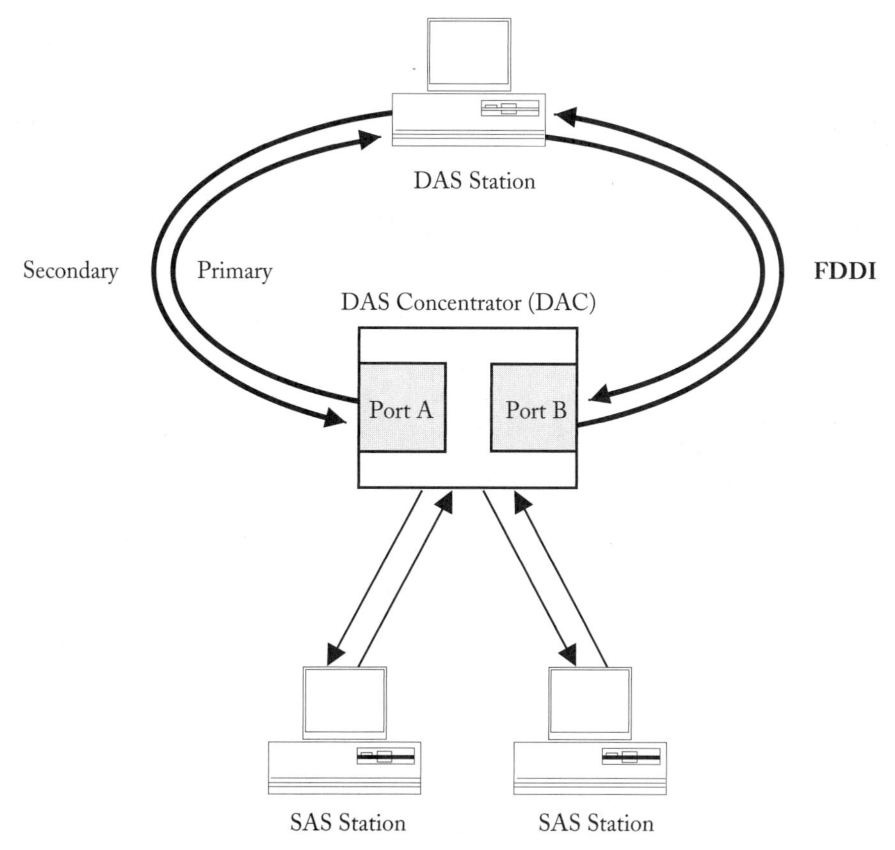

Figure 7.12 *Single Attachment Stations, Dual Attachment Stations, and Concentrators*

When a single device fails, devices on either side of the failed device wrap, forming a single ring. See Figure 7.14. The remaining devices on the ring are unaffected. When a cable failure occurs, devices on either side of the cable fault wrap. See Figure 7.13. Again, the remaining devices on the ring are unaffected. FDDI can only provide fault tolerance against a single failure. If two or more failures occur, the wrap function will cause the FDDI ring to be segmented into two or more separate and isolated rings.

Optical Bypass Switch

The optical bypass switch is optional and is used to provide continuous dual-ring operation if a device on the dual ring fails. This is used to avoid ring segmentation and isolate failed devices from the ring. If a failure of a DAS device occurs, such as a loss of power, the optical bypass switch will pass the light signal through itself, thereby maintaining the integrity of the ring. Optical bypass switches are not inexpensive. The purpose of using them is to avoid a wrapped ring condition in case of a device failure. Figure 7.15 shows an optical bypass switch in an FDDI network.

Dual Homing

Mission-critical devices, such as routers, servers, or mainframe hosts, can use another fault-tolerant technique called dual homing to provide additional redundancy. To provide dual homing the mission-critical device is attached to a pair of concentrators. Figure 7.16 illustrates dual-homed servers and routers.

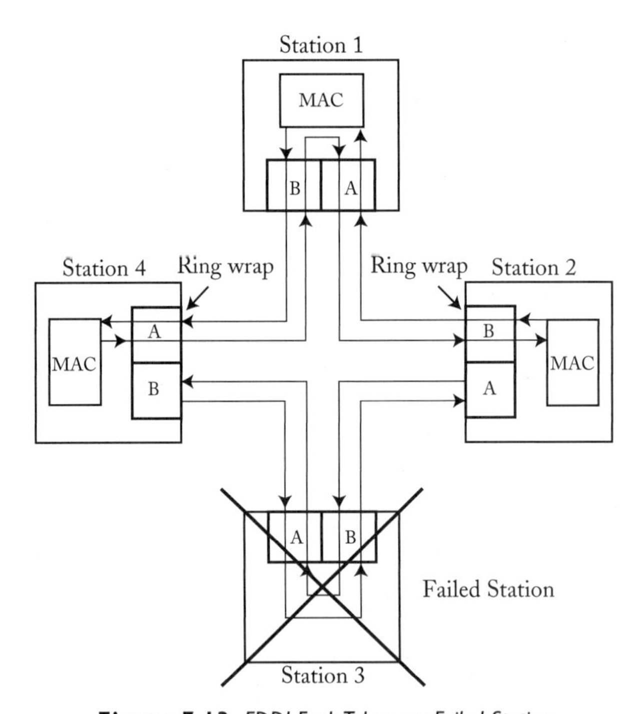

Figure 7.13 *FDDI Fault Tolerance: Failed Station*

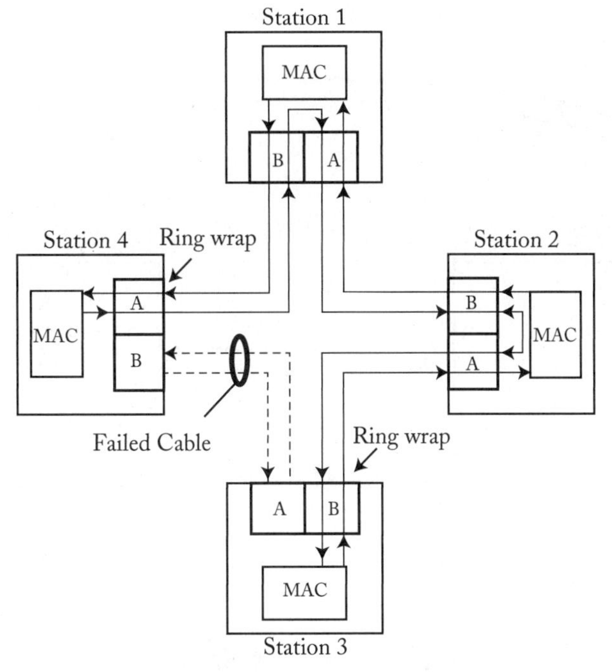

Figure 7.14 *FDDI Fault Tolerance: Failed Cable*

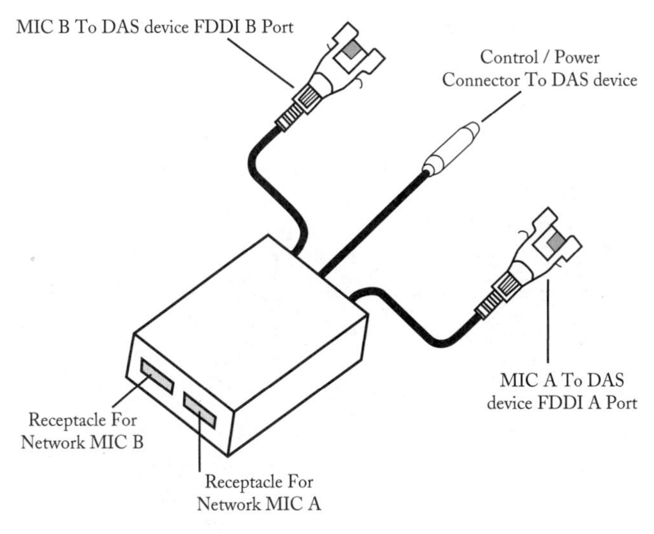

Figure 7.15 *FDDI Optical Bypass Switch*

One concentrator link from each device is the active link; the other concentrator link is passive. As with the primary and secondary FDDI rings, the passive link remains in backup mode until the primary link (or its concentrator) is determined to have failed. When a failure occurs, the passive link takes over.

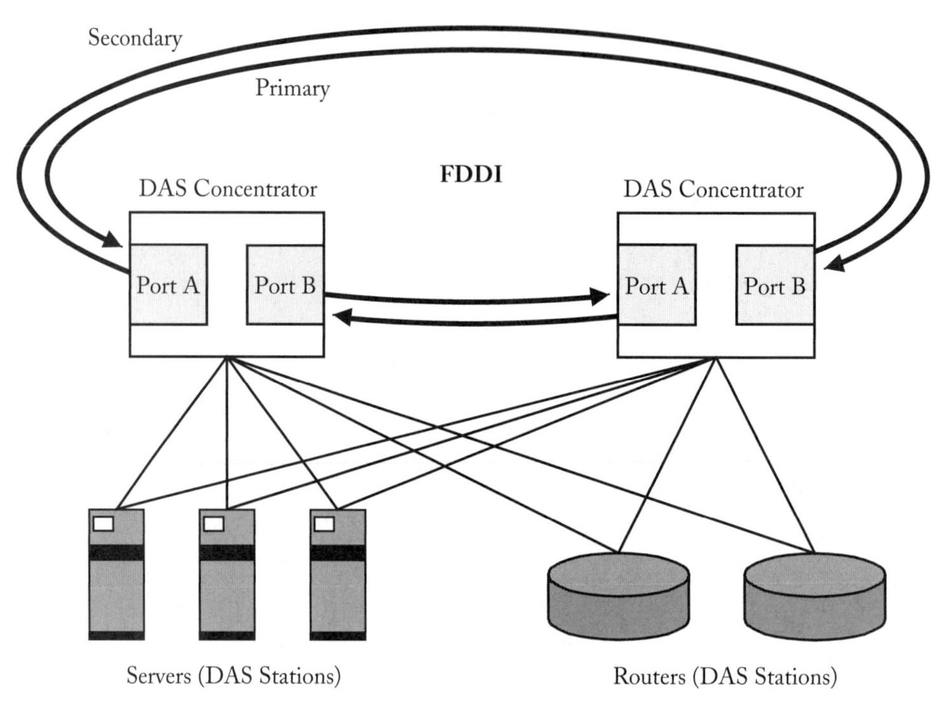

Figure 7.16 *FDDI/CDDI Dual-Homed Servers and Routers*

CHAPTER SUMMARY

Since their creation in the mid-1980s, both token ring and FDDI evolved to support UTP cabling in addition to STP and fiber. As with Ethernet, when new cabling standards emerged, new token ring and CDDI specifications were adapted to support them. Embracing these new cabling standards yielded many benefits, perhaps most significant being the radical simplification of token ring design rules. In addition to added media flexibility, features and functionality have been built into the LAN technologies to ensure reliable service, and in the case of FDDI, fault-tolerant service. Unfortunately, the engineering of token-passing rings has been anything but elegant or simple; the products remain expensive and the network designs overly complex. As a result, these technologies have outlived their usefulness and are no longer being used for new LAN installations.

REVIEW QUESTIONS

1. Identify two differences between token ring and CDDI.
2. Identify two differences between FDDI and CDDI.
3. What is the purpose of the phantom drive?
4. Which LAN uses the phantom drive?
5. What is a token?
6. ANSI X3T12 operates at what data rate?
7. List three types of media supported by FDDI.
8. Define the following acronyms:

FDDI	CDDI
DAS	DAC
SAS	MIC
SMT	TP-PMD
CAU	LAM

9. What is the difference between a MSAU, CAU, and LAM?
10. What is the purpose of Dual Homing?

Contemporary LAN Standards

OBJECTIVES

After completing this chapter, you should be able to:

- Identify the key differences between classic Ethernet and Fast Ethernet.
- Identify the key differences between Fast and Gigabit Ethernet.
- Define the MAC and PHY.
- Explain the purpose of FLP and auto-negotiation.
- Explain the difference between binary, ternary, and quinary.
- Describe the various types of multi-mode and single-mode fiber.
- Describe the differences between half duplex and full duplex.
- Define MII and GMII.
- Define the role and functions of repeaters.
- Explain Carrier Extension and Frame Bursting.

INTRODUCTION

Learn about the specifications and media used by the latest LAN technologies: Fast Ethernet and Gigabit Ethernet. Also part of the body of work known as IEEE 802.3, these standards are known as 100Base-T and 1000Base-T. Most of the new standards for 802.3 Ethernet rely on media standards defined by ANSI/TIA/EIA, including the various categories of UTP cabling, components, installation practices, and certification criteria.

IEEE 802.3u 100BASE-T FAST ETHERNET

IEEE 802.3u defines several implementations of 100 Mbps Ethernet and is commonly referred to as Fast Ethernet. Based on 10Base-T Ethernet, Fast Ethernet uses the same frame structure, and supports CSMA/CD as well as full-duplex operation. The term *100Base-X* refers to two of the most common Fast Ethernet implementations: 100Base-FX and 100Base-TX. Two other Fast Ethernet standards virtually never used include 100Base-T4 and 100Base-T2.

The standards actually define the implementation of 100 Mbps Fast Ethernet over several different kinds of media as defined by ANSI/EIA/TIA, including two pairs of Category 5 UTP, four pairs of Category 3 UTP, two pairs of Category 3 UTP, two pairs of 150 Ohm STP, and duplex multi-mode optical fiber compliant with ISO/IEC 11801.

ANSI/EIA/TIA cabling standards define the media as well as the cable plant in terms of megahertz (MHz), whereas ISO/IEC ANSI/IEEE networking standards define network technologies in terms of megabits per second (Mbps). Rather than define unique cable plant requirements for its network technologies, more recent ISO/IEC ANSI/IEEE networking standards, such as several of the Fast Ethernet and Gigabit Ethernet specifications, simply refer to the ANSI/EIA/TIA cabling standards.

In addition to defining the electrical properties of the cable and components, ANSI/EIA/TIA standards also define cable plant installation practices and certification requirements. ANSI/EIA/TIA cabling standards also define the performance criteria for the entire communications channel, from one end of a cable run to the other. However, some ISO/IEC ANSI/IEEE networking standards impose more stringent requirements on communications channels than those established by the ANSI/EIA/TIA cabling standards. Although most of these "special needs" networking standards never made it into production, the effort expended in their development has sometimes contributed to later advancements in networking, resulting in even more advanced standards. Each of the Fast Ethernet standards are shown in Table 8.1.

100BASE-TX

The original Fast Ethernet signaling scheme supports 100 Mbps operation over two pairs of Category 5 UTP. Rather than re-creating the wheel, the 100Base-TX com-

TABLE 8.1 IEEE 802.3u 100 MBPS "FAST ETHERNET"

Sub-committee	*Channel Length*	*Media*
100Base-TX	100 meters (328 feet)	2-pair Cat5 UTP
100Base-FX	412 meters (1312 feet)	Half-duplex, 2 multi-mode optical fibers
	2000 meters (6560 feet)	Full-duplex, 2 multi-mode optical fibers
100Base-T4	100 meters (328 feet)	4-pair Cat3 UTP or better
100Base-T2	100 meters (328 feet)	2-pair Cat3, Cat4, or Cat5 UTP

mittee borrowed its 100 Mbps transmission scheme from the FDDI twisted pair Physical Medium Dependent Interface (TP-PMD). Commonly referred to as CDDI, it uses 4B/5B data encoding with MLT-3. 100Base-TX supports 100-meter cable lengths and is capable of full-duplex operation. 100Base-TX is by far the most popular of all the Fast Ethernet implementations. 100Base-TX specifies the use of 8-position modular RJ-45 connectors (see Figure 8.1).

The 100Base-TX standard also supports 150 Ohm STP cabling. This is the same cabling defined for use with IEEE 802.5 token ring, and uses DB-9 connectors. The STP implementation of Fast Ethernet will be ignored since it is so rare (if not entirely nonexistent).

100BASE-FX

Also an original Fast Ethernet signaling scheme, 100Base-FX is quite popular. It supports 100 Mbps operation over two multi-mode optical fibers using 4B/5B encoding. Because 100Base-X leverages the Physical layer standards of ISO/IEC 9314 and ANSI X3T12 (FDDI), it also supports the media defined by those standards. 100Base-FX relies on the optical fiber specifications defined in ISO/IEC 9314-3:1990, which includes 62.5/125 µm duplex multi-mode fiber (MMF) and a wavelength of 1300 nanometers. The standard also supports other types of MMF, including 50/125 µm,

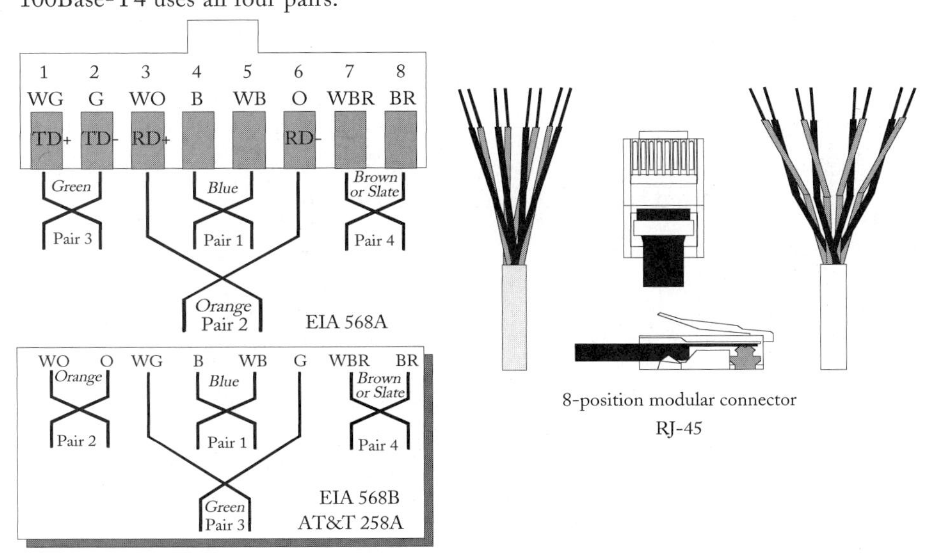

Figure 8.1 *IEEE 802.3 Fast Ethernet UTP Cable and Connectors*

85/125 μm, and 100/140 μm, although they may not support the same Link Segment lengths as 62.5/125 μm fiber.

Optical fiber sizes are expressed in micrometers (μm). The first number represents the diameter of the fiber core while the second number represents the diameter of the cladding material that surrounds the core. Light wavelengths are expressed in nanometers (nm). The visible light spectrum consists of a range of wavelengths from approximately 700 nm to approximately 400 nm (or 7×10^{-7} m to 4×10^{-7} m).

100Base-FX supports 412-meter fiber cable lengths in half-duplex operation, and up to 2 km in full-duplex operation, using 62.5/125 μm fiber. 100Base-FX recommends the use of duplex fiber SC connectors, but the FDDI MIC is also permitted (see Figure 8.2).

100BASE-T4

This variation was developed at the same time as 100Base-TX. It is an alternative 100 Mbps signaling scheme designed to support 100 Mbps operation over Category 3 UTP, a media which enjoyed a rather large installed base at the time. Unlike 100Base-TX, 100Base-T4 requires all four-pairs to support 100-meter cable lengths. As with 10Base-T and 100Base-TX, 100Base-T4 relies on the eight-position modular (RJ-45) connector.

100Base-T4 uses the 8B/6T (ternary) encoding scheme to convert eight data bits into six ternary symbols for transmission over three wire pairs. In contrast to binary encoding, which uses two values, 0 and 1, ternary signals may have one of three values: −1, 0, or +1. Each octet of data is mapped to a pattern of six ternary symbols, called a 6T code group. The 6T code groups are fanned out to three independent serial channels. This scheme effectively distributes the 100 Mbps data rate over three pairs that can each carry 33.333 Mbps in one direction at a time. The ternary symbol transmission rate on each pair is 6/8 times 33.333 Mbps, or 25 megabaud. (To calculate this, simply realize that $6/8 = 3/4$ or .75, and then multiply $.75 \times 33.333 = 24.999$.) This results in a maximum spectral bandwidth of 12.5 MHz, which is well within the 16 MHz range supported by each pair of Cat3 UTP.

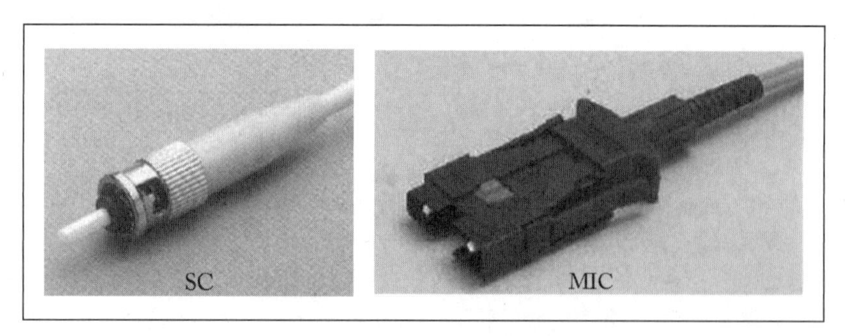

Figure 8.2 *IEEE 802.3 Fast Ethernet UTP Optical Fiber Connectors*

Of the four pairs used by 100Base-T4, one pair is dedicated to transmit data, one pair is dedicated to receive data, and two bi-directional pairs are used to either transmit or receive data. This ensures that one pair is always available to allow collisions to be detected on the link, while three pairs are available to carry data. Because of this, 100Base-T4 is not capable of full-duplex operation. Use of 100Base-T4 requires unique transceiver (PHY) hardware, and relies on the ever-popular RJ-45 connector (see Figure 8.1).

In addition, although 100Base-T4 is intended to support Cat3 UTP (or better), it imposes additional certification requirements on the media—so many, in fact, that it is almost an oxymoron to consider it a standard cable plant. And businesses, it seems, did not view upgrading their cable plants to Category 5 UTP to be an insurmountable hurdle or excessive expense, therefore 100Base-T4 was never widely deployed.

100BASE-T2

A newer 100 Mbps signaling scheme that supports 100 Mbps operation over two pairs of Category 3 UTP, 100Base-T2 supports 100-meter cable lengths and is capable of full-duplex operation. Again, since most businesses have already upgraded their cable plants to Category 5 UTP, there has been little demand for 100Base-T2. This technology does, however, embody several interesting technological developments.

100Base-T2 uses a complex data-encoding scheme called "PAM 5x5." It is a five-level, five-phase (hence 5×5) constellation-space signaling method. Data is transmitted using a "quinary" (five-level) signal that can have the following values: −2, −1, 0, +1, or +2. Four bits of information are transmitted per signal transition on both wire pairs with a transition rate of 25 megabaud (4 × 25 = 100). 100Base-T2 employs a dual duplex transmission technique where both pairs transmit data in both directions simultaneously. Hybrids and cancellers are used to remove the resulting crosstalk of the transmitted signals from received signals. The system is extremely complex and would not have been practical even a few years ago. However, with the availability of low-cost, 0.6 micron silicon technology, the large amount of circuitry required for the complex Digital Signal Processing (DSP) is possible in a corner of the PHY chip.

And as with 100Base-T4, 100Base-T2 is intended to support Cat3, Cat4, and Cat5 UTP, but it imposes numerous additional certification requirements on the media. Also like 100Base-T4, the standard has yet to be widely deployed since its approval.

FAST ETHERNET FEATURES

Along with higher data rates, Fast Ethernet introduced several additional features and capabilities that classic Ethernet lacks. Auto-negotiation was developed to ensure backward compatibility between the original 10 Mbps Ethernet and the new 100 Mbps Fast Ethernet components. Although some 10 Mbps Ethernet gear has supported full-duplex operation for some time, it did so without the benefit of standardization. Full-duplex operation was formally included in the Fast Ethernet standards.

Auto-Negotiation

The Fast Ethernet standards include an auto-negotiation protocol intended to determine the speed and duplex used by the devices at each end of a Link Segment (i.e., the performance options common to both link partners). The auto-negotiation protocol relies on a signaling method called the Fast Link Pulse (FLP). As each Fast Ethernet transceiver transmits its own FLP over its Link Segment, it is listening for the FLP coming from the device at the other end. Auto-negotiation works its way through a sequence of performance options to determine the highest common denominator between the link partners.

While a great idea in theory, in practice the implementation of auto-negotiation has been less than perfect. The FLP function is implemented in the Physical layer (PHY) chipset. Various manufacturers of Fast Ethernet PHY chipsets apparently implement the FLP function differently enough to cause some incompatibilities. The result is two interconnected devices selecting different performance parameters.

For example, consider a Category 3 UTP Link Segment with a 10/100 Ethernet hub at one end and a 10/100 Ethernet DTE/station at the other end. Both devices support auto-negotiation, but a hub cannot support full duplex. When the hub and station are powered up, auto-negotiation will attempt to determine the performance capabilities common to both devices. The highest performance mode they have in common is 100Base-TX, half duplex. Since auto-negotiation FLPs are simply grouped bursts of the same link pulses used in 10Base-T, the pulses will travel over Category 3 UTP cabling without any problems. However, the two devices will probably select 100Base-TX half-duplex operation, which requires Category 5 UTP cabling. This results in either no data being transmitted or poor performance due to high bit error rates, even thought the link status indicator is illuminated on the devices at both ends of the Link Segment. The link status LEDs indicate only circuit continuity over a given Link Segment they do not ensure proper electrical properties for reliable data transmission. And since auto-negotiation cannot detect the category of cabling being used, the network installer is responsible for providing the correct cabling to support 100 Mbps operation (see Figure 8.3).

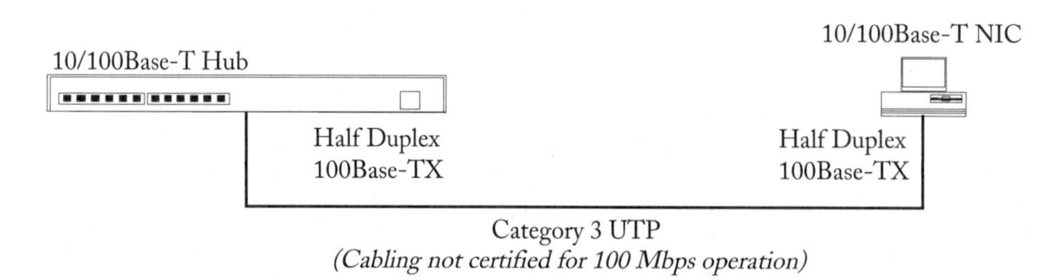

Figure 8.3 *Auto-Negotiation: Problems Determining Appropriate Speed for Media*

Another, more common problem involves the correct selection of full- or half-duplex operation. In this scenario, a Category 5 UTP Link Segment connects a 10/100 Ethernet switch at one end and a 10/100 Ethernet DTE/station at the other end. Both devices correctly conclude that their partner device supports 100 Mbps operation, but they fail to arrive at the same conclusion for duplex operation. If one device is operating in full duplex while its partner device is operating in half duplex (employing CSMA/CD), many transmission errors will be generated. This is because the first device permits simultaneous transmission and reception of data, whereas the second device considers this a collision. Not only is this problem very common, it can be very difficult to isolate and resolve (see Figure 8.4).

Auto-negotiation is a common problem between different manufacturers. However, it can also be a problem within a single manufacturer's product line if that manufacturer uses different PHY chipsets across its product line. As odd as this may sound, it has been known to happen.

With the exception of 100Base-T4, all Fast Ethernet implementations are capable of supporting full-duplex operation. And it should be noted that the Category 5 UTP cabling standards do not support full-duplex operation, but we do it anyway. Be aware that your results may vary. Enabling full duplex often causes no ill effects, while in some cases network throughput may drop significantly. This is often due to high error rates caused by excessive return loss and/or near-end crosstalk (NEXT) problems. These problems may be caused by a number of factors throughout the cable plant, such as mismatched components, poor punch-downs, or patch cords that are not up to spec (which is frequently the case).

As one final point, because the FLP are simply grouped bursts of the same link pulses used in 10Base-T, a 100Base-TX device can auto-negotiate with an ordinary 10Base-T device. While many newer 10Base-T NICs will support full-duplex operation, many older products will not. And if two devices fail to auto-negotiate properly, it is sometimes possible to manipulate the parameters in either the NIC driver's utility program or via the administrative interface of the hub, switch, or router. Smaller hubs, switches, and routers usually lack this ability to manually control the PHY chip registers.

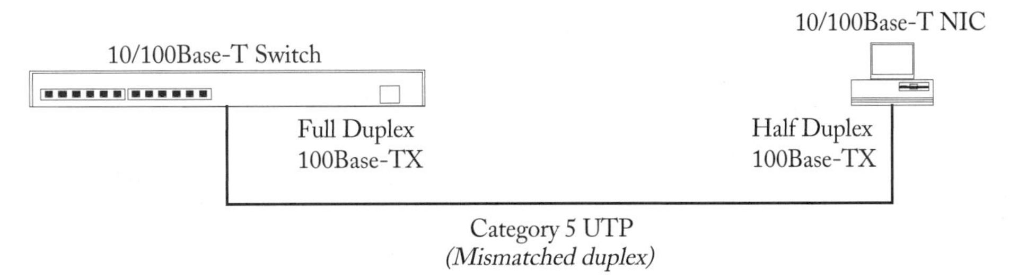

Figure 8.4 *Auto-Negotiation: Problems Determining Correct Duplex*

Full-Duplex Transmission

Full duplex requires switches and cannot be used with hubs. When using switches, there is no shared physical medium, therefore contention is eliminated. When full-duplex mode is enabled, CSMA/CD is essentially disabled. Because there can be no contention for access to the media, there is no need to sense carrier or defer to the presence of another transmission. Concurrent transmission and reception of data is expected with full duplex, therefore Collision Detection must be disabled.

The standard provides that the Physical layer interface may indicate to the MAC layer that there are concurrent transmissions by both stations on a given Link Segment, but since these transmissions do not interfere with each other a MAC operating in full-duplex mode must not respond to such transmissions as if they represented a collision. Full-duplex stations do not defer to received traffic, abort transmission, issue Jam signals, execute the back-off algorithm, and reschedule transmissions as part of the normal CSMA/CD protocol would require. A station configured to support full duplex may transmit whenever a frame is queued. A full-duplex Link Segment can support the full data rate of a given LAN technology in both directions simultaneously, e.g., a 100Base-TX Link Segment can support 100 Mbps in both directions, virtually doubling the effective bandwidth of the Link Segment. Note that it does not increase the effective throughput in either direction beyond the technology's 100 Mbps data rate.

INSTALLATION RULES

IEEE 802.3 Fast Ethernet standards specify support for TIA/EIA 568-A standard cabling. As such, the Fast Ethernet standards do not provide as much detail on the engineering of the actual media or cable plant design. With few exceptions, the IEEE 802.3 Fast Ethernet standards defer to the TIA/EIA 568-A cabling standards. IEEE 802.3 does define several electrical (and optical) media characteristics necessary to ensure proper operation of each of the Fast Ethernet standards. Any cable plant intended to support Fast Ethernet should be tested and certified to meet or exceed the characteristics specified for that particular media (e.g., Cat5 or optical fiber).

Repeaters are an integral part of all Fast Ethernet networks consisting of more than two DTE/stations. Two stations may be directly connected using a specially wired crossover cable called a Null Link Segment. If the network is to consist of more than two stations, a repeater is required to interconnect them. As with classic Ethernet, normal Link Segments are wired straight through (see Figure 8.5).

Repeaters extend the network topology by joining two or more Link Segments. All stations interconnected by repeaters are in the same Collision Domain. In some cases, a single Collision Domain may incorporate multiple repeaters to provide the maximum connection path length. Whereas classic Ethernet supports up to four (4) repeater hops, Fast Ethernet is limited to just one or two repeaters in a given Collision Domain depending on the class of repeater employed.

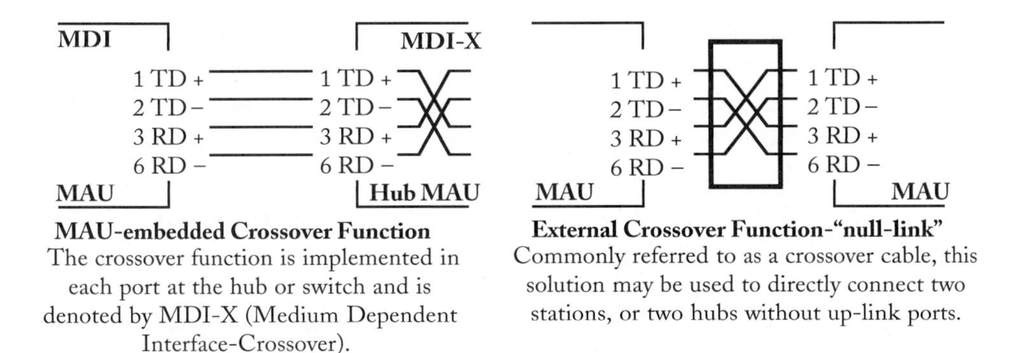

Figure 8.5 *100Base-TX Fast Ethernet MDI and MDI-X Crossover Function*

The Fast Ethernet standards specify two classes of repeater: Class I and Class II. Class I repeaters support just one repeater per Collision Domain, and Class II repeaters support a pair of repeaters per Collision Domain. With a maximum Link Segment of 100 meters to each DTE/station, the two Class II repeaters may be separated by no more than 5 meters.

To overcome this limitation, many Fast Ethernet hubs support a *stackable* configuration. Stackable hubs use a short proprietary cable to link multiple stacked hubs. Because the proprietary cable is so short, the hubs must be stacked directly on top of each other for the cables to reach the special connectors that are usually provided on the back of each of the hubs. Stackable hubs from different vendors typically support anywhere from four to eight hubs in a single stack. And because the stacking feature is proprietary, hubs from different vendors cannot be mixed.

If larger network topologies are required, switches and/or routers will be required to interconnect multiple Collision Domains. Topics such as switches, routers, and internetworking are beyond the scope of this chapter and will be addressed later in this book.

IEEE 802.3z AND 802.3ab GIGABIT ETHERNET

IEEE 802.3z defines several implementations of 1000 Mbps Ethernet and is commonly referred to as Gigabit Ethernet, or Gig-E. Based on 10Base-T Ethernet and 100Base-TX Fast Ethernet, Gigabit Ethernet uses the same frame structure, and supports both CSMA/CD and full-duplex operation. Since Gigabit Ethernet is intended to provide high-capacity connections, full duplex is preferred. There are few, if any, implementations of actual Gig-E products that rely on CSMA/CD.

The term 1000Base-X refers collectively to the IEEE 802.3z 1000Base-SX, 1000Base-LX, and 1000Base-CX implementations of Gigabit Ethernet. A separate development effort called IEEE 802.3ab 1000Base-T is responsible for the long-haul copper implementation that supports 100 meters of Cat5 UTP. The standards actually define the implementation of 1000 Mbps Gigabit Ethernet over several different kinds of media as defined by ANSI/EIA/TIA, including four pairs of Category 5 UTP (compliant with ISO/IEC11801), duplex multi-mode optical fiber, and duplex single-mode fiber (compliant with IEC 60793-2:1992).

IEEE 802.3z 1000Base-X borrows the physical signaling technology from the ANSI X3.230-1994 Fibre Channel standard (FC-0 and FC-1). 1000Base-X uses 8B/10B transmission encoding, which converts data bytes into DC-balanced bit streams, and provides integrated clock recovery and increased error detection. With 8B/10B data encoding, every eight data bits are converted into a 10-bit symbol for transmission over the media. The resulting overhead of the extra bits requires a signal transmission rate of 1.25 gigabaud to deliver a net throughput of 1 Gbps of user data.

This is the same encoding scheme used by ANSI X3.230-1994 (FC-PH) standard for Fibre Channel. 1000Base-X supports 1000Base-LX long-wavelength single-mode fiber (SMF), 1000Base-SX short-wavelength multi-mode fiber (MMF), and a short-haul copper patch cord named 1000Base-CX (which defines a 150-Ohm balanced two-pair shielded cable). Each implementation can support half-duplex (CSMA/CD) or full-duplex operation.

1000Base-X also defines a Gigabit Media Independent Interface (GMII) that interfaces between the Media Access Control (MAC) and PHY functions of a Gigabit Ethernet device. GMII is analogous to the Attachment Unit Interface (AUI) in 10 Mbps Ethernet, and the Media Independent Interface (MII) in 100 Mbps Ethernet. However, unlike the AUI and MII, no actual connector or cable is defined for GMII. The 1000Base-X standard requires that the transceiver function must be built into all Gigabit Ethernet devices. It is impractical to expose the GMII as an external interface due to the high 125 MHz frequency of its parallel 8-bit data transfers. Therefore the GMII exists only as an internal component interface.

IEEE 802.3ab 1000Base-T does not use these same Fibre Channel standards and is not part of IEEE 802.3z. 1000Base-T was engineered to support Cat5 UTP; however, several problems emerged with the Cat5 media, prompting ANSI/TIA/EIA to tighten the Cat5 certification requirements and re-release the cabling specifications as Category 5 Enhanced, or Cat5E.

To further complicate matters, TIA independently opted to develop another Gigabit Ethernet specification called 1000Base-TX (PN-4657) that requires just two pairs of Category 6 UTP. Whereas Cat5/Cat5E supports a spectral bandwidth of 100 MHz, Cat6 supports 250 MHz, thereby providing more headroom. However, the TIA 1000Base-TX specification supports 1000 Mbps over just two pairs, which consumes much of that newly found headroom. And although several different standards already exist for Cat6 UTP, the electrical properties of the channel have not yet been fully defined. This means that there are no certifications to which a Cat6 cable plant can adhere. Until there are, no cable plant can be certified as Cat6 compliant.

While everyone agrees that Gigabit Ethernet will run over UTP, there are significant differences in how these various standards bodies are approaching the problems. For example, the IEEE 802.3ab 1000Base-T specifications rely on more expensive electronics to support cheaper and more common Cat5 cable, while TIA's 1000Base-TX draft proposal relies on the newer and more expensive Cat6 cabling to support cheaper electronics. In essence, the IEEE places the burden on the digital electronics side, relying on advanced Digital Signal Processing (DSP) hardware, while the TIA places the burden on the analog cabling side, relying on advanced cabling technology.

Regardless of the potential benefits, the IEEE has yet to accept the TIA's 1000Base-TX proposal, and it is not being supported by any of the networking equipment manufacturers. The essential difference is in the requirement for echo cancellation in the four-pair, bi-directional IEEE approach, and dealing with the faster frequencies and Electro-Magnetic Interference (EMI) issues on the 2×2-pair unidirectional TIA approach. The separate Gigabit Ethernet standards are shown in Table 8.2.

1000BASE-LX

The "L" in 1000Base-LX refers to the "long wavelength" LASERs used to transmit signals over duplex optical fiber. It supports 1000 Mbps operation over two multi-mode optical fibers, or two single-mode optical fibers (SMF) using the same 8B/10B encoding defined for use with Fibre Channel, and supports similar optical and electrical specifications.

1000Base-LX was designed to support long-haul campus backbone applications. Support is specified for 50 μm MMF, 62.5 μm MMF, and 10 μm SMF using wavelengths

TABLE 8.2 IEEE 802.3Z/802.3AB 1000 MBPS GIGABIT ETHERNET

Sub-committee	*Channel Length*	*Media*
1000Base-LX	316 meters (1036 feet)	MMF or SMF, half-duplex
	550 meters (1804 feet)	MMF, full-duplex
	5000 meters (16,404 feet)	SMF, full-duplex
1000Base-SX	275 meters (902 feet)	62.5/125 μm MMF, half-duplex
	316 meters (1036 feet)	50/125 μm MMF, half-duplex
	275 meters (902 feet)	62.5/125 μm MMF, full-duplex
	550 meters (1804 feet)	50/125 μm MMF, full-duplex
1000Base-CX	25 meters (82 feet)	2-pair balanced 150-Ohm STP, or twinax
1000Base-T	100 meters (328 feet)	4-pair Cat5 UTP or better
1000Base-TX	100 meters (328 feet)	2-pair Cat6 UTP [TIA DRAFT]

in the range from 1270 to 1355 nanometers. Obtaining the maximum Link Segment length depends on the use of full-duplex transmissions and the type of optical fiber used. 1000Base-LX specifies the use of duplex SC connectors.

1000BASE-SX

The "S" in 1000Base-SX refers to the "short wavelength" LASERs used to transmit signals over duplex optical fiber. 1000Base-SX supports 1000 Mbps operation only over two multi-mode optical fibers, again using the same 8B/10B encoding defined for use with Fibre Channel, supporting similar optical and electrical specifications.

1000Base-SX was designed to support horizontal cabling runs and shorter backbone applications. Support is specified for 50 μm MMF and 62.5 μm MMF using wavelengths in the range from 770 to 860 nanometers. The 1000Base-SX standard does not support single-mode fiber. As with the 1000Base-LX standard, obtaining the maximum Link Segment length depends on the use of full-duplex transmissions and the type of optical fiber used. 1000Base-SX also specifies the use of duplex SC connectors (see Figure 8.2).

1000BASE-CX

The "C" in 1000Base-CX refers to "copper" and it uses specially shielded balanced copper jumper cables or "short haul copper." Segment lengths are limited to just 25 meters (82 feet), which relegates 1000Base-CX for use as as a patch cord to connect equipment in small areas like wiring closets. 1000Base-CX specifies either the 9-pin shielded D-subminiature connector, or 8-pin ANSI Fibre Channel Type 2 (HSSC) connector (see Figure 8.6).

Some cabling vendors are working to develop enhanced copper media and connector technology to improve its performance and support greater distances. This enhanced media is referred to by such terms as twinax and quad cabling. These cables may consist of individually shielded wire pairs, or a single overall shield surrounding all wire pairs.

Figure 8.6 *1000Base-CX HSSC Connector*

1000BASE-T

Formally released in June 1999, this long-haul copper Gigabit Ethernet signaling scheme supports 1000 Mbps operation over four pairs of Cat5 UTP, and a unique physical signaling technology developed just for Gigabit Ethernet. 1000Base-T employs advanced multilevel signaling over four pairs of Category 5 balanced copper cabling. The 1000Base-T standards working group borrowed from the signaling methods developed for both 100Base-TX (125 Megabaud transmissions), 100Base-T (multi-pair transmission) and 100Base-T2 (PAM5x5 encoding). 1000Base-T uses 8B1Q4 data encoding and a new transmission scheme referred to as four-dimensional, five-level pulse amplitude modulation (4D-PAM5).

In 4D-PAM5 the transmitted symbols are selected from a "four-dimensional five-level symbol constellation." The four dimensions refer to the four pairs used in each 1000Base-T Link Segment. Data is transmitted over each of the four pairs using a "quinary" (five-level) signal consisting of the following values: −2, −1, 0, +1, or +2. Through this encoding scheme, eight data bits are converted to one transmission of four quinary symbols.

1000Base-T employs a bi-directional dual duplex transmission technique where all four pairs transmit data in both directions simultaneously (no mean feat!). Each conductor of each wire pair carries 250 Mbps at the same time. Advanced DSP is used to remove the resulting crosstalk of the transmitted signal from a received signal.

The 1000Base-T Physical Medium Attachment (PMA) sublayer includes several convenience features, such as the ability to automatically detect and correct for pair-swapping and accidental crossover connections, and the ability to automatically detect and correct reversed polarity in the pair connections. Because 1000Base-T simultaneously transmits across four wire pairs, the PMA also includes the ability to automatically correct for differential delay variations (or signal skew) across the wire pairs.

1000Base-T supports 100-meter cable lengths Cat5 or Cat5E and the ubiquitous eight-position modular (RJ-45) connector (see Figure 8.1). Although the standards support both half-duplex (CSMA/CD) and full-duplex operation, full duplex is always preferred. By comparison, 100Base-TX Fast Ethernet can also support full-duplex operation, but it transmits on one pair and receives on a separate pair.

GIGABIT ETHERNET FEATURES

Along with an even higher data rate, Gigabit Ethernet introduced several features and capabilities to address the special requirements of 1000 Mbps data transmissions. The addition of Carrier Extension (CE) was required to support CSMA/CD over 100-meter Link Segments at 1000 Mbps. Frame Bursting was added to overcome the increased overhead created by CE and improve overall performance.

Carrier Extension

Because of the effect of the high data rate with CSMA/CD, half-duplex implementations require the use of CE to support 100-meter Link Segments. At operating speeds faster than 100 Mbps, the standard SlotTime of 512 bit times becomes inadequate to accommodate network topologies based on 100-meter UTP Link Segments. CE provides a means by which the SlotTime duration can be sufficiently increased to support the desired topologies without increasing the minimum frame size.

The alternative would be to change the minimum frame size for Gigabit Ethernet, but this would cause incompatibilities with other Ethernet implementations. Therefore, CE relies on the use of non-data bits, referred to as extension bits, which are appended to frames that are less than the desired SlotTime length. In so doing, the resulting transmissions are extended to at least the duration of the new SlotTime. The maximum length of the new SlotTime with CE is 4096 bits.

The MAC layer monitors the medium for collisions while it is transmitting CE bits, and it will treat any collision that occurs after the SlotTime threshold as a late collision. Without CE, CSMA/CD timing would restrict the Link Segments to about 20 meters. With full-duplex transmission, CE is unnecessary. In summary: just forget half-duplex Gig-E.

Frame Bursting

At operating speeds faster than 100 Mbps such as Gigabit Ethernet, an option is provided whereby a given implementation may transmit multiple successive frames without relinquishing control of the transmission medium. This mode of operation is referred to as burst mode and is supported only in half-duplex implementations. The purpose of Frame Bursting is, at least in part, to compensate for the inefficiencies of CE.

In essence, once a station has successfully transmitted one frame, the transmitting station can begin transmission of another frame without contending for the medium because all of the other stations on the network are already deferring to it. The transmitting station must fill the InterFrame Spacing interval with extension bits. Extension bits are readily distinguished from data bits at the receiving stations, which will continue to detect carrier as long as a bit stream appears on the medium. This allows the transmitting station to initiate subsequent frame transmissions until a specified burst limit is reached. The burst limit value is 65,536 bits (64kb).

The first frame of a burst will be extended if necessary. Subsequent frames within a burst do not require extension. In a properly configured, error-free network, collisions cannot occur once the first frame of a burst, plus any extension bits, has been transmitted. If the MAC layer does detect a collision after the first frame of a burst, or after the SlotTime has been reached, that collision will be treated as a late collision. The presence of late collisions indicates an out-of-spec network topology.

INSTALLATION RULES

As with Fast Ethernet, the IEEE 802.3 Gigabit Ethernet standards also specify support for TIA/EIA 568-A standard cabling. As such, the Gigabit Ethernet standards do not provide much detail on the engineering of the actual media or cable plant design, and instead defer to the TIA/EIA 568-A cabling standards. IEEE 802.3 does define several electrical (and optical) media characteristics necessary to ensure proper operation of each of the Gigabit Ethernet standards. Any cable plant intended to support Gigabit Ethernet should be tested and certified to meet or exceed the characteristics specified for that particular media (e.g., Cat5, Cat5E, Cat6 or optical fiber).

Repeaters are an integral part of all Gigabit Ethernet networks consisting of more than two DTE/stations. Two stations may be directly connected using a specially wired Gigabit Ethernet crossover cable. If the network is to consist of more than two stations, a repeater is required to interconnect them. As with classic Ethernet and Fast Ethernet, normal Link Segments are wired straight through; however, 1000Base-T Gigabit Ethernet uses all four pairs (see Figure 8.7).

As with classic and Fast Ethernet, repeaters extend the network topology by joining two or more Link Segments, and all stations interconnected by repeaters are in the same Collision Domain. All Gigabit Ethernet Link Segments are connected directly to a single repeater. There is no provision for multiple repeaters in Gigabit Ethernet. Only one Gigabit Ethernet repeater may exist between any two DTE/stations within a single Collision Domain.

1000Base-T Crossover Cable Pinout (MDI-X)

Side 1 Pin (Signal)	Side 2 Pin (Signal)
1 (TP0+)	3 (TP1+)
2 (TP0-)	6 (TP1-)
3 (TP1+)	1 (TP0+)
6 (TP1-)	2 (TP1-)
4 (TP2+)	7 (TP3+)
5 (TP2-)	8 (TP3-)
7 (TP3+)	4 (TP2+)
8 (TP3-)	5 (TP2-)

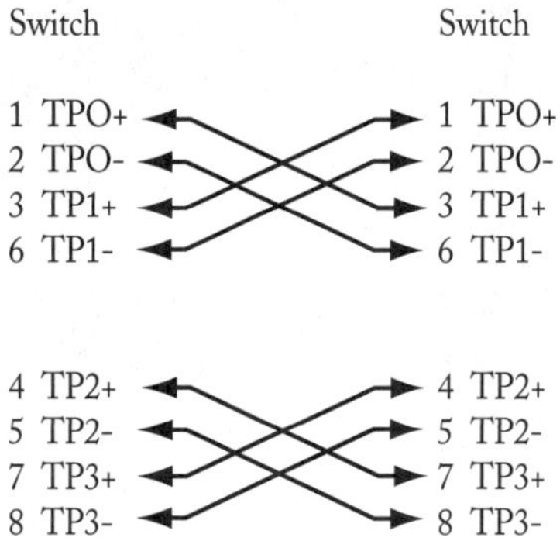

Switch	Switch
1 TPO+	1 TPO+
2 TPO-	2 TPO-
3 TP1+	3 TP1+
6 TP1-	6 TP1-
4 TP2+	4 TP2+
5 TP2-	5 TP2-
7 TP3+	7 TP3+
8 TP3-	8 TP3-

Figure 8.7 *1000Base-T Gigabit Ethernet MDI and MDI-X Crossover Cable Pinout*

If larger network topologies are required, switches and/or routers will be required to interconnect multiple Collision Domains. Topics such as switches, routers, and internetworking are beyond the scope of this chapter and will be addressed later in this book.

CHAPTER SUMMARY

The incredible market success of 100 Mbps Fast Ethernet expedited the development of Gigabit Ethernet. By the time you read this, the 10 Gigabit Ethernet standards will have been approved, and faster versions are already in the works. However, Fast and Gigabit Ethernet will likely satisfy most business requirements for quite some time. Faster technologies will be relegated to equipment closets and server farms until such time higher speeds can be properly exploited at the desktop.

Today's LANs of choice are 100Base-TX, 100Base-FX, and some 1000Base-T. 1000Base-SX/LX are good choices for high-speed backbone interconnections between buildings on a campus, or between floors in larger buildings. Although there are few if any desktop computers that can fill a 1000 Mbps pipe today, many high-end servers certainly can. The task for network installers is to select the appropriate LAN technology for each situation, and, as always, balance price and performance. Fortunately the upgrade path from 10 Mbps Ethernet to 100 Mbps Fast Ethernet to 1000 Mbps Gigabit Ethernet is clearly marked and easy to follow (as well as economical)!

Obviously these newer, high-speed LAN standards are very involved and their underlying technologies can be quite complicated. It should be noted that this text by no means covers all of the specifications detailed in the standards documents, just that information deemed useful to understand the networks. The information provided herein is adequate for those responsible for the installation and maintenance of such networks, not for those interested in designing or manufacturing components.

REVIEW QUESTIONS

1. Identify three differences between Ethernet and Fast Ethernet.
2. Identify three differences between Fast Ethernet and Gigabit Ethernet.
3. Describe the purpose and shortcomings of auto-negotiation.
4. Which LAN standards use 4B/5B, 8B/10B, and 8B1Q4?

5. What is the difference between binary, ternary, and quinary?
6. List two types of MMF.
7. What is the difference between Class I and Class II repeaters?
8. Define the following acronyms:

MAU	MSAU
MAC	PHY
MII	GMII
EMI	NEXT
MMF	SMF

9. Identify two advantages of full-duplex operation.
10. What is the purpose of Carrier Extension and Frame Bursting?

Internetworking Devices: Repeaters and Hubs

OBJECTIVES

After completing this chapter, you should be able to:

- Define the functions of a repeater.
- Explain the relationship of repeaters to the OSI model.
- Identify the differences between repeaters at 10 Mbps, 100 Mbps, and 1000 Mbps.
- Explain the repeater rules for 10 Mbps, 100 Mbps, and 1000 Mbps Ethernet.
- Explain the meaning of SQE and identify potential problems it can cause.
- Define the Collision Domain for 10 Mbps, 100 Mbps, and 1000 Mbps Ethernet.
- Define the meaning of SlotTime and describe the round-trip propagation delay.
- Explain the differences between Class I and Class II repeaters.
- Explain the differences between a repeater and a buffered distributor.
- Define the timing parameters for 10 Mbps, 100 Mbps, and 1000 Mbps Ethernet.

INTRODUCTION

This chapter will address the most basic components of network interconnection: repeaters. Also known as hubs and concentrators, these devices enable LANs to be extended to the maximum distance allowed by each network topology. As a basis for comparison, all devices will be defined according to their relationship to the ISO OSI seven-layer reference model. This will help to provide an understanding of the function of each device and the relationship to other devices.

AN OVERVIEW OF REPEATERS

The simplest of LAN extension components are repeaters. Also known as hubs and sometimes concentrators, repeaters are used to extend the physical distance of a given local area network (LAN) topology. By joining multiple LAN segments, repeaters allow for larger LAN topologies and a greater number of DTE/stations. In fact, the only way to support the maximum number of 1024 stations permitted in a single Ethernet LAN is by using repeaters.

Repeaters operate at the bottom of the OSI model at layer 1, the Physical layer. A repeater forwards all bit streams received on any interface to all other interfaces on that repeater. Repeaters retime and regenerate each bit stream without regard to MAC layer addressing, upper-layer addressing, or errors within the MAC frame. Since repeaters require only the ability to read transmitted signals at the Physical layer (i.e., binary), they are very fast and reliable, and impose very little latency. The function of a repeater is transparent to all DTE/stations on the LAN (see Figure 9.1).

LAN repeaters are local devices only—they do not support wide area network (WAN) connectivity. Repeaters can support greater distances using optical fiber media. Repeaters are the digital equivalent of an (analog) amplifier. But unlike an amplifier, repeaters do not regenerate the noise accumulated along the transmission path. Repeaters do regenerate each received bit stream to the original amplitude and restore bit timing to its original condition.

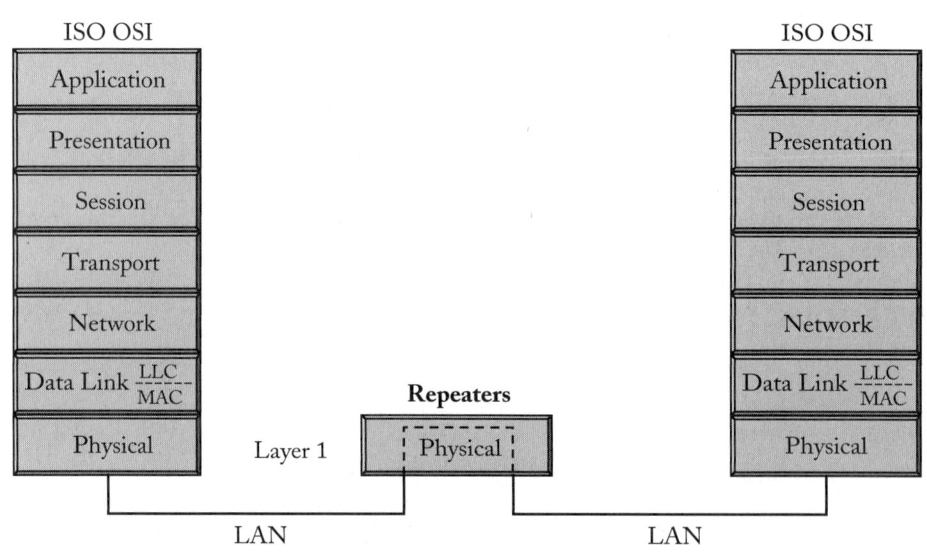

Figure 9.1 *Repeaters and the OSI Model*
Repeaters correspond to the functions defined in ISO OSI Physical layer (layer 1).

Ethernet repeaters forward all digital signals on the wire, including valid frames, misaligned frames, corrupt frames, frames with a bad CRC value, excessively long frames, short frames, fragments, and runts. Fragments are always interpreted as the result of a collision and are forwarded as a 96-bit Jam signal to reinforce the collision event across the entire Collision Domain.

If a collision occurs, the repeater must propagate the collision event to all network segments by transmitting a Jam signal. Repeaters must also protect the network by isolating the carrier activity from faulty Link Segments thus preventing the propagation of errors through the network. This is performed by a function called auto-partitioning. For example, if a repeater port detects over thirty sequential collisions on any port, that port will be automatically partitioned from the rest of the network. When the fault is cleared and a valid frame appears on that port, the auto-reconnection feature will allow the Link Segment to rejoin the network.

Repeaters do not implement the full CSMA/CD protocol, at least not in the same way as DTE/stations. For example, rather than employing the back-off algorithm after a collision is detected as all DTE/stations do, repeaters are required to transmit a Jam signal to ensure that all DTE/stations see the collision event.

IEEE 802.3 transceivers support an optional feature called the Signal Quality Error (SQE) test. A transceiver with SQE enabled will assert a signal over the Collision In pair of the AUI cable after it successfully transmits each frame. IEEE 802.3–compliant DTE/stations correctly interpret this signal; however, repeaters do not. When a repeater sees the SQE signal, it is interpreted as a collision and the repeater transmits Jam signals on all other ports. This behavior is undesirable since it has deleterious effects on network performance. It is therefore necessary to disable SQE on all transceivers connected to repeaters. This problem typically occurs only with external transceivers.

Multiport repeaters are also available for other network technologies such as token ring and FDDI. Multiport token ring repeaters are sometimes referred to as "active MSAUs." Small, two-port token ring repeaters are called "lobe extenders" and are used to extend the length of a single token ring lobe cable beyond its nominal distance. Multiport repeaters for FDDI and CDDI are commonly referred to as concentrators. Since Ethernet in its various incarnations is by far the dominant LAN technology in the world, it will be the focus of the remainder of this chapter.

Redundant paths or loops are never permitted between repeaters. If there is a requirement for fault-tolerant connections, a backup path, or larger network topologies than those supported with repeaters alone, then switches and/or routers must be used to interconnect multiple Collision Domains. Topics such as switches, routers, and internetworking are beyond the scope of this chapter and will be addressed later in this book.

THE ETHERNET COLLISION DOMAIN

As discussed previously, Ethernet uses a network access method called carrier sense multiple access with Collision Detection (CSMA/CD). A collision occurs when two or more devices transmit at the same time. Collisions are normal events in Ethernet, and Collision Detection provides a mechanism to deal with them after the fact.

All Ethernet LAN segments interconnected by repeaters become one Collision Domain. This means that all collisions, as well as all traffic, occurring on any one LAN segment will be propagated to the other LAN segments. All DTE/stations within a given Collision Domain must share the capacity of that LAN technology, whether it is 10, 100, or 1000 Mbps. Because CSMA/CD Ethernet defines a shared network environment, collisions can occur between any two (or more) DTE/stations anywhere within a Collision Domain. Collisions must be detected within the time it takes to complete the transmission of the smallest valid Ethernet frame (64 bytes or 512 bits). For 10 Mbps Ethernet, each bit time is 100 ns; therefore, the time required to transmit a 64-byte frame is 51.2 µs (512 bits × 100 ns = 51.2 µs). The time translates into distance, and defines the SlotTime used to calculate the round-trip collision propagation delay for a given LAN.

Therefore, the maximum extent of a given Ethernet LAN (including all media, transceivers, and repeaters) is the time required for a transmitted frame to traverse the entire LAN, encounter a collision, and propagate that collision event back to the originating DTE/station while that station is still transmitting the frame. This can take no longer than the SlotTime defined for that Ethernet LAN.

Each of the IEEE 802.3/Ethernet standards defines a set of parameterized values that govern all transmissions. Table 9.1 gives the values for 10, 100, and 1000 Mbps Ethernet.

TABLE 9.1 PARAMETERIZED VALUES OF 10/100/1000 MBPS ETHERNET

Parameters	*10 Mbps*	*100 Mbps*	*1000 Mbps*
Bit time (BT)	100 ns	10 ns	1 ns
SlotTime	512 BT (51.2 µs)	512 BT (5.12 µs)	4096 BT (4.096 µs)
InterFrame Gap	96 BT (9.6 µs)	96 BT (.96 µs)	96 BT (.096 µs)
Minimum frame	64 bytes	64 bytes	64 bytes
Maximum frame	1518 bytes	1518 bytes	1518 bytes
Burst limit	NA	NA	65,536 bits

10 MBPS REPEATERS

The packet forwarding delay across a 10 Mbps repeater is typically no more than 8 bit times. For 10 Mbps Ethernet, the delay equates to 800 nanoseconds (8×100ns). The delay imposed by repeaters is negligible; however, all forms of media add their own signal propagation velocity delay. Virtually all media includes a performance characteristic referred to as its nominal velocity of propagation (NVP), which is expressed as a percentage of the speed or velocity of light (c). For example, most UTP cabling exhibits an NVP of approximately 0.59c. Stated another way, electrical signals travel through UTP cabling at about 59% of the speed of light (see Figure 9.2).

Copper Ethernet repeaters are available in a wide variety of configurations. Some offer only two interfaces, or ports, to interconnect two segments of the same or different media. A repeater must be used whenever different media require interconnection. Today even the smallest 10Base-T repeaters provide several RJ-45 ports. Since 10Base-T supports only one DTE/station per port, a two-port repeater would not be very practical. These miniature 10Base-T hubs, mini-hubs or "hublets" as they are sometimes called, are in fact multiport repeaters and should be implemented with care so as to not violate the repeater and segment rules. The smallest hubs generally provide no management capabilities, whereas the midsize and large enterprise-level hubs usually do provide management capabilities based on the simple network management protocol (SNMP) used in TCP/IP networks. In some cases, vendors will provide proprietary management solutions.

	Media Segment Type	Maximum Number of MAUs per Segment	Maximum Segment Length (meters)	Minimum Media Propagation Velocity*	Maximum Media Delay per Segment (nanoseconds)
Coax					
	10Base5	100	500	0.77c	2165
	10Base2	30	185	0.65c	950
Link					
	FOIRL	2	1000	0.66c	5000
	10Base-T	2	100†	0.59c	1000
	10Base-FL	2	2000	0.77c	2165
	AUI‡	1 DTE/1 MAU	50	0.65c	257

* $c = 3 \times 10^8$ m/s (speed of light)
† Actual maximum link segment length depends on cabling characteristics.
‡ AUI is not a segment.

Figure 9.2 *Cabling and Nominal Velocity of Propagation Delay*
The NVP delay affects the maximum extent of all segments regardless of repeaters.

The rules limiting the use of multiple or cascaded repeaters are relatively simple. The rules vary based on the speed of the LAN. The rules for 10 Mbps Ethernet do not apply to 100 Mbps Fast Ethernet, neither of which apply to 1000 Mbps Gigabit Ethernet. The 10 Mbps Ethernet repeater implementation specifications are sometimes referred to as the "5, 4, 3, 2, 1" rules. Stations in a 10 Mbps Ethernet Collision Domain may not traverse more than five segments, four repeaters to communicate with any other station within that Collision Domain. Three of those segments may be populated with stations, two are inter-repeater links (IRL); the sum of which defines one Collision Domain (see Figure 9.3).

10 Mbps Ethernet Repeater Rules

5 segments maximum, interconnected by
4 repeaters (or hubs) in the transmission path
3 of the segments in the path may be populated with stations
2 or more are inter-repeater Link Segments (for distance)
1 Collision Domain

10BASE2 ETHERNET REPEATERS

10Base2 repeaters correspond to the repeater rules illustrated in Figure 9.3. Most 10Base2 repeaters are multiport repeaters, and most provide 50-Ohm termination internally on each BNC, or bayonet connector. This eliminates the need for a BNC "T" connector and BNC terminator at each interface port on the hub. A 50-Ohm terminating resistor is still required at the other end of the coaxial segment.

To confirm that the port is in fact terminated internally, use a cable scanner or simple multimeter to measure the impedance at one of the interface ports on the multiport repeater. First measure the impedance directly; if the meter reports 50 Ohms (±2%), the repeater port is internally terminated. If a different impedance value is reported, the repeater is probably not providing internal termination.

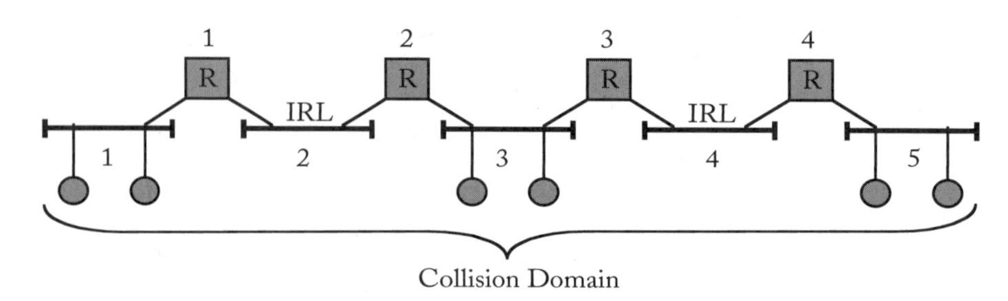

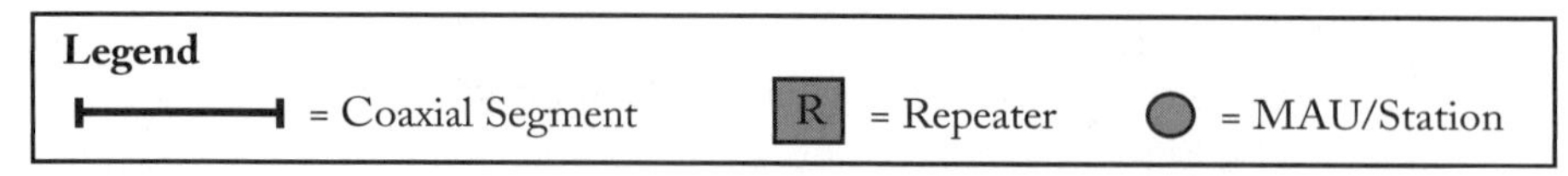

Figure 9.3 *IEEE 802.3 Ethernet Repeater Rules and Collision Domain*
This diagram illustrates the maximum diameter of a 10Base5 or 10Base2 Ethernet Collision Domain.

For the next test, connect a section of 10Base2 coax directly to a port on the repeater, install one BNC "T" and one 50-Ohm terminator to the other end of the coaxial cable, and measure the impedance at the open "T" connector. If it reads 25 Ohms, the repeater is providing internal termination. This may be expressed as $\left(\frac{r^1 + r^2}{r^1 \times r^2}\right)$.

Add a second BNC "T" and a second terminator and take the measurement again. It will read 16 Ohms. This impedance is incorrect for proper Ethernet operation, so remove the extra terminator. Keep the terminator at the far end of the coaxial segment.

10BASE-T ETHERNET REPEATERS

Better known as hubs, all 10Base-T repeaters are multiport repeaters. Repeaters are an integral part of all 10Base-T LANs consisting of more than two DTE/stations. Repeaters extend the network topology by joining two or more Link Segments. All Link Segments interconnected by repeaters are said to be in the same Collision Domain. A 10 Mbps Ethernet Collision Domain may incorporate multiple repeaters to provide the maximum connection path length. The Link Segment from each DTE/station is connected directly to a 10Base-T repeater/hub (see Figure 9.4).

As with all repeaters, 10Base-T Ethernet repeaters are not considered a DTE/station and do not count toward the overall limit of 1024 stations within a single Collision Domain. A repeater must be able to receive and decode data from any Link Segment under the worst-case noise, timing, and signal amplitude conditions. All bit streams received on any repeater interface are retransmitted onto all other Link Segments attached to it with timing, amplitude, and coding restored. Data retransmission occurs simultaneously as the bit stream is being received with a delay of no more than 8 bit times (800 ns).

Figure 9.4 provides an example of the maximum extent achievable in a 10 Mbps Ethernet Collision Domain using 10Base-T and UTP media. These examples illustrate

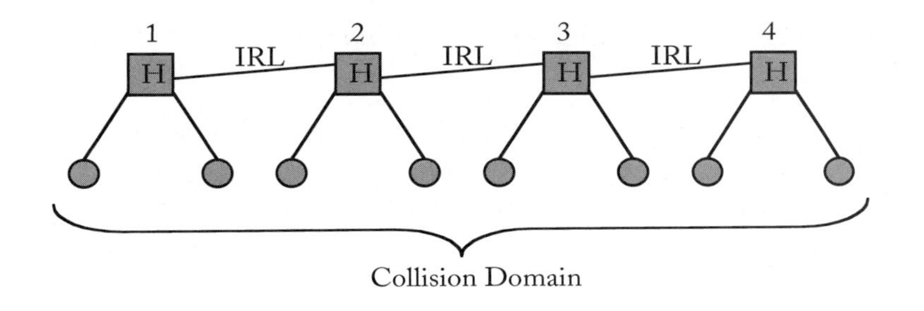

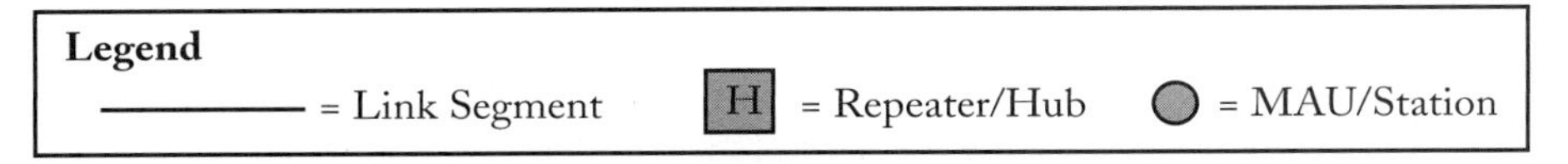

Figure 9.4 *10Base-T Ethernet Repeaters*
This diagram illustrates the maximum diameter of a 10Base-T Ethernet Collision Domain.

the simplest of linear configurations and do not cover tree topologies. The intent is to define the maximum permissible data path between the farthest stations for each of the various standard media. Additional repeaters may be added as long as there are no more than four repeaters and five segments between the farthest stations. The best way to accomplish this is to restructure the entire network into two hierarchical tiers.

Repeaters are an integral part of all Fast Ethernet networks consisting of more than two DTE/stations. Repeaters extend the network topology by joining two or more Link Segments. All Link Segments interconnected by repeaters are said to be in the same Collision Domain. In some cases, a single Collision Domain may incorporate multiple repeaters to provide the maximum connection path length. In most applications, Link Segments are connected directly to a single repeater; however, the standards also support the use of a pair of repeaters connected by an inter-repeater link (IRL) segment.

Fast Ethernet repeaters are not considered a DTE/station and do not count toward the overall limit of 1024 stations within a single Collision Domain. A repeater must be able to receive and decode data from any Link Segment under the worst-case noise, timing, and signal amplitude conditions. All bit streams received on any repeater interface are retransmitted onto all other Link Segments attached to it with timing, amplitude, and coding restored. Data retransmission occurs almost simultaneously as the bit stream is being received.

100 MBPS FAST ETHERNET REPEATERS

As with other Ethernet repeaters, if a collision occurs the repeater must propagate the collision event throughout the network by transmitting a Jam signal. Repeaters must also protect the network by isolating the carrier activity from faulty Link Segments thereby preventing it. This is performed by a function called auto-partitioning. When the fault is cleared, auto-reconnection allows the Link Segment to rejoin the network.

IEEE 802.3u Fast Ethernet defines two classes of repeaters:

Class I
Only one such repeater may exist between any two DTE/stations within a single Collision Domain. For example, if using Cat5 UTP, a Class I repeater will support multiple 100 meter Link Segments for a Collision Domain with a maximum diameter of 200 meters. If using 100Base-FX optical fiber, a Class I repeater will support multiple 412 meter Link Segments for a Collision Domain with a maximum diameter of 824 meters (see Figure 9.5).

Class II
Only two such repeaters may exist between any two DTE/stations within a single Collision Domain. The two Class II repeaters may be connected by an inter-repeater link of no more than 5 meters. For example, if using Cat5 UTP, a pair of Class II

repeaters will support multiple 100 meter Link Segments for a Collision Domain with a maximum diameter of 205 meters. If using 100Base-FX optical fiber, a pair of Class II repeaters will support multiple 412 meter Link Segments for a Collision Domain with a maximum diameter of 829 meters (see Figure 9.6).

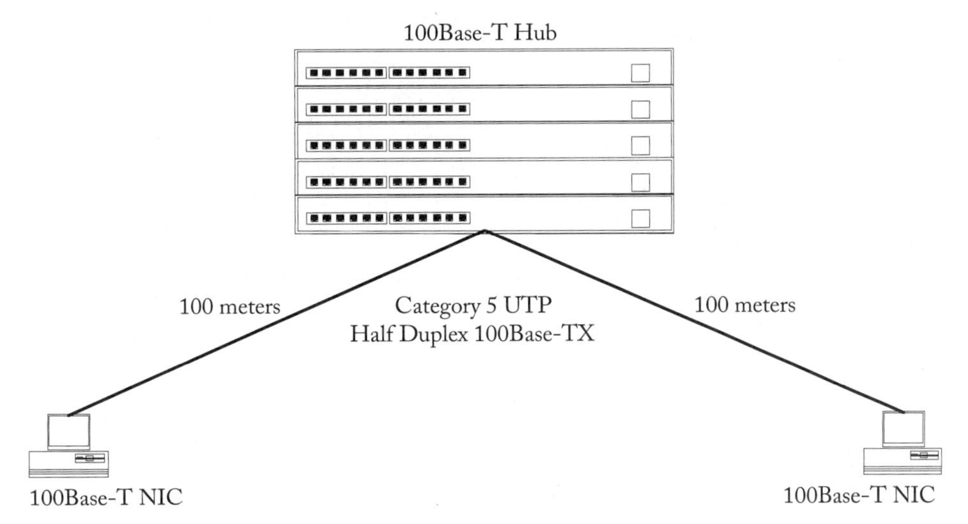

Figure 9.5 *Fast Ethernet Repeaters: Class I*

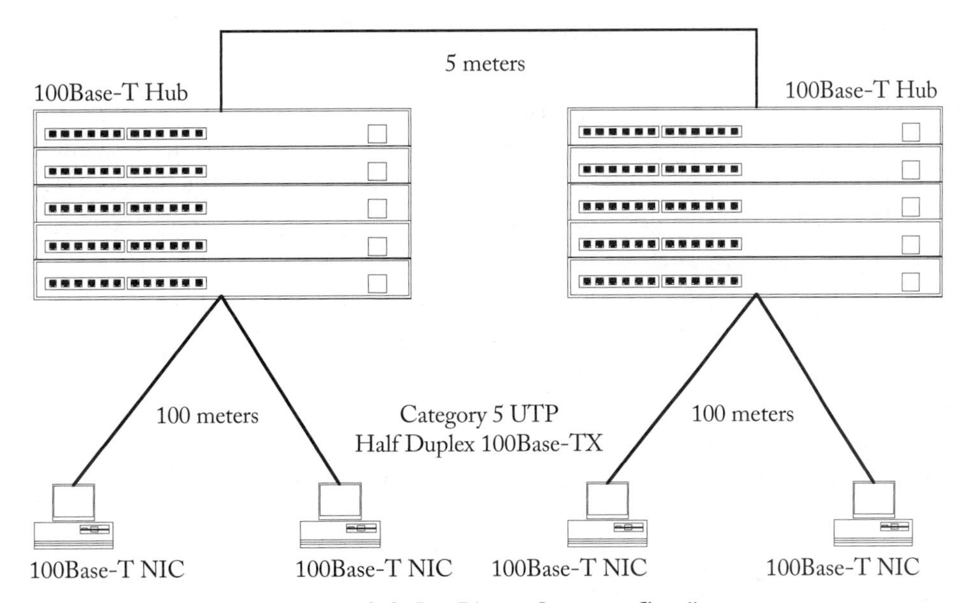

Figure 9.6 *Fast Ethernet Repeaters: Class II*

1000 MBPS GIGABIT ETHERNET REPEATERS

Repeaters are an integral part of all Gigabit Ethernet networks consisting of more than two DTE/stations. As before, repeaters extend the network topology by joining two or more Link Segments. All Link Segments interconnected by repeaters are said to be in the same Collision Domain. All Gigabit Ethernet Link Segments are connected directly to a single repeater. There is no provision for multiple repeaters in Gigabit Ethernet.

Only one Gigabit Ethernet repeater may exist between any two DTE/stations within a single Collision Domain. For example, if using Cat5 UTP, a Gig-E repeater will support multiple 100 meter Link Segments for a Collision Domain with a maximum diameter of 200 meters. If using 1000Base-SX with multi-mode optical fiber, a Gig-E repeater will support multiple 275 (or 316) meter Link Segments for a Collision Domain with a maximum diameter of 550 (or 632) meters (depending on the type of MMF used). And if using 1000Base-LX with single-mode fiber, a Gig-E repeater can support multiple 550 meter Link Segments for a Collision Domain with a maximum diameter of 1100 meters (see Figure 9.7).

Gigabit Ethernet repeaters are not considered a DTE/station and do not count toward the overall limit of 1024 stations within a single Collision Domain. A repeater must be able to receive and decode data from any Link Segment under the worst-case noise, timing, and signal amplitude conditions. All bit streams received on any repeater interface are retransmitted onto all other Link Segments attached to it with timing, amplitude, and coding restored. Data retransmission occurs almost simultaneously as the bit stream is being received.

As with other Ethernet repeaters, if a collision occurs the repeater must propagate the collision event throughout the network by transmitting a Jam signal. Repeaters must also protect the network by isolating the carrier activity from faulty Link Segments thereby preventing it. This is performed by a function called auto-partitioning. When the fault is cleared, auto-reconnection allows the Link Segment to rejoin the network.

BUFFERED REPEATERS

Traditional 10 Mbps Ethernet and 100 Mbps Fast Ethernet must be configured as either a shared network (using repeaters) or dedicated network (using switches). Gigabit Ethernet can provide a third option: the "buffered repeater" or "buffered distributor." The purpose of such devices is to strike a balance between the higher performance of switched Ethernet and the lower cost of shared Ethernet. Buffered repeaters function somewhere in between those two technologies.

Specifically, Gigabit Ethernet buffered distributors share 1 Gbps of network capacity among all Gig-E devices. By contrast, a switch dedicates a full 1 Gbps to each Gig-E device connected to it. Like a switch, buffered distributors provide full-duplex connections to each attached Gig-E device. By content, all Ethernet repeater links are

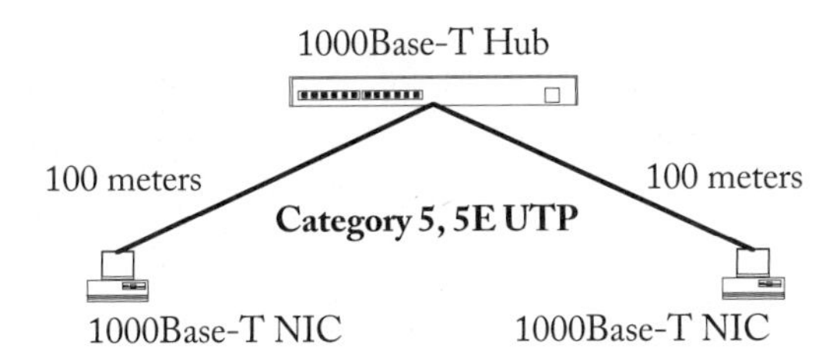

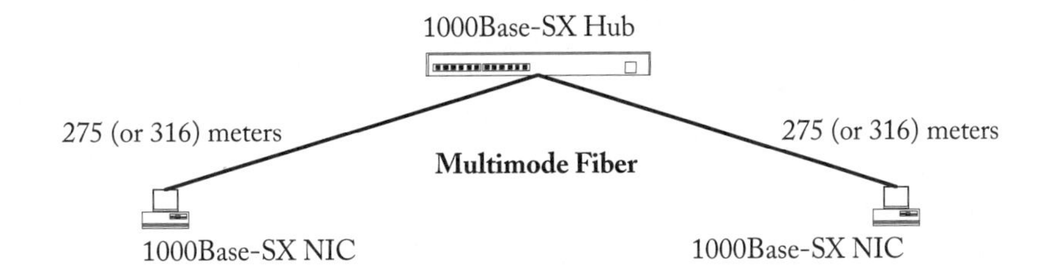

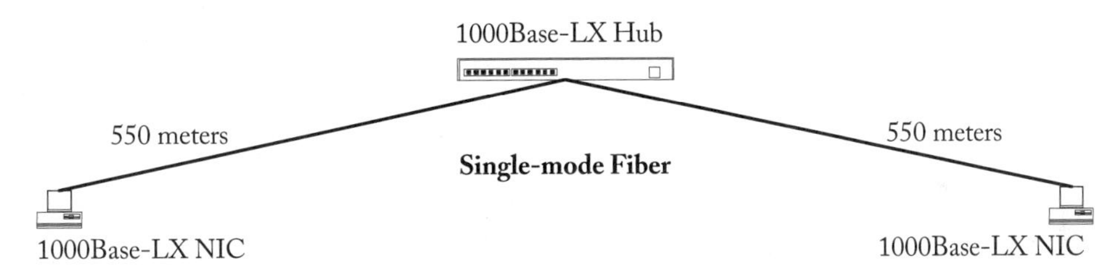

Figure 9.7 *Gigabit Ethernet Repeaters*

exclusively half-duplex. As the name implies, when congestion exists in the network a buffered distributor buffers incoming frames and holds those frames for transmission until the opportunity to transmit arises (see Figure 9.8).

The benefit of using such hybrid devices is in the ability to support full-duplex links. Although it may seem that full-duplex links in a half-duplex network may be a zero-sum gain, in this case there is a real benefit. The use of full-duplex Ethernet at any speed eliminates collisions because stations are permitted to send and receive packets at the same time. It has also been observed that the most significant performance gains of full duplex are realized in high-speed networks, such as Gigabit Ethernet.

Because buffered distributors provide a first-in, first-out (FIFO) buffer for each Gig-E interface, Collision Domain timing issues can be avoided, thereby obviating the associated distance restrictions. Frame-based flow control based on the IEEE 802.3x standard is used to prevent FIFO buffers from overflowing. By providing full-duplex connections to every attached Gig-E device, the buffered distributor can deliver significantly higher performance than a simple half-duplex repeater. Furthermore, the price of buffered distributors will probably remain significantly lower than switches, making them an attractive alternative to half-duplex repeaters and full-duplex switches.

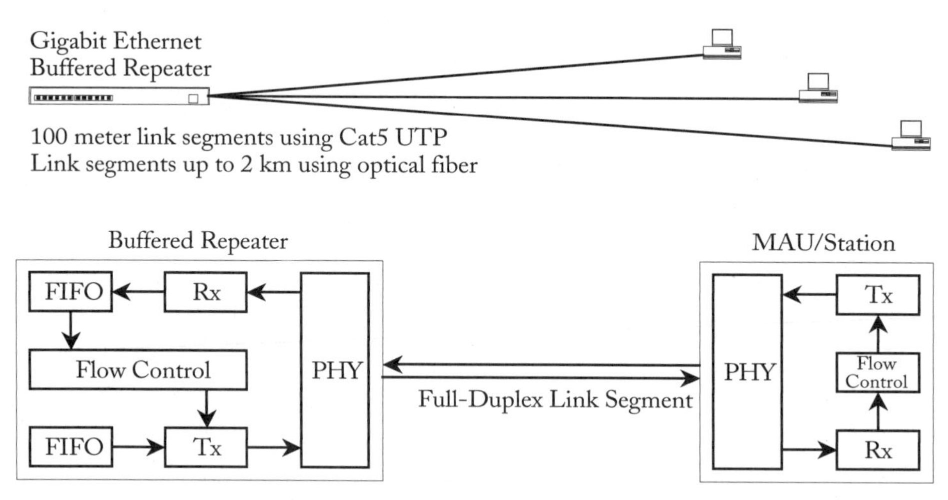

Figure 9.8 *Gigabit Ethernet Buffered Repeaters*

CHAPTER SUMMARY

Repeaters are the simplest of network interconnection devices. They are used to extend the physical topology of networks to achieve greater distances and support more stations. Today's multiport repeaters are called hubs and range in capacity from very small, two- or four-port units, to midsized units supporting 8 to 24 ports, to large rack-mounted chassis supporting more than 100 interfaces.

Repeaters are inexpensive and simple to use; however, they do have several drawbacks. Because repeater-based Ethernet networks rely on CSMA/CD they are always restricted to half-duplex operation. Also, whereas switch-based networks provide dedicated bandwidth to each station attached to them, all stations connected to a repeater must share the throughput capacity of that repeater. In a repeater-based network, only one transmission can take place at any given moment. The one exception to this is the buffered distributors used in Gigabit Ethernet.

In conclusion, layer 2 switches are rapidly displacing repeaters. Switches provide many benefits including superior performance and management capabilities, and the difference in cost is becoming negligible. Furthermore, providers of applications such as Voice over IP (VoIP) are mandating switch-based networks and will no longer support repeater-based networks. Put simply, repeaters are out and switches are in.

REVIEW QUESTIONS

1. Define the functions of a repeater.
2. Explain the relationship of repeaters to the OSI model.
3. Identify three differences between 10 Mbps repeaters and 100 Mbps repeaters.
4. Identify three unique features of 1000 Mbps repeaters.
5. Explain the "5, 4, 3, 2, 1" Ethernet repeater rules.
6. Define the Ethernet Collision Domain for 10, 100, and 1000 Mbps Ethernet.
7. What is SQE and what can be done to avoid potential problems with it?
8. Define the following acronyms:

SQE	NVP
µm	nm
Mbps	Gbps
Gig-E	FIFO
BNC	BT

9. What is the difference between a repeater and a buffered distributor?
10. What is the SlotTime value for 10, 100, and 1000 Mbps Ethernet?

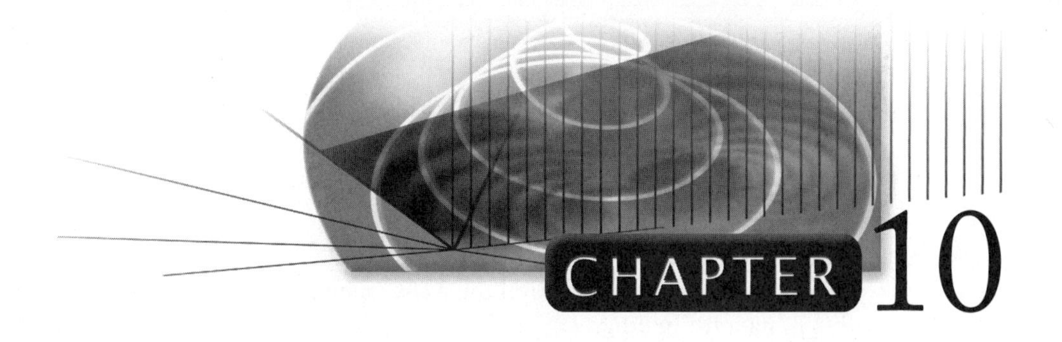

Internetworking Devices: Bridges and Layer 2 Switches

OBJECTIVES

After completing this chapter, you should be able to:

- Explain the relationship of bridges and switches to the OSI model.
- Identify the differences between bridges and repeaters.
- Explain the purpose and process of route discovery.
- Identify applications for source route bridging.
- Define translation and encapsulation bridging.
- Define the functions of 802.1D bridging.
- Identify applications for transparent bridging.
- Explain the differences between bridges and switches.
- Explain the differences between circuit, cell, and frame switches.
- Define the purpose, functions, and problems of Spanning Tree Protocol.

INTRODUCTION

This chapter will address the most common components of network interconnection: bridges and layer 2 switches. These devices enable LANs to be extended well beyond the maximum distance allowed by their respective network topologies. Again, as a basis for comparison all devices will be defined according to their relationship to the ISO OSI seven-layer reference model. This will help to provide an understanding of the function of each device and the relationship to other devices.

AN OVERVIEW OF BRIDGES

There are several varieties of bridges designed to solve different problems and support different network technologies. Virtually all devices called bridges are referred to as layer

2 devices because they fulfill the functions defined by layer 2 of the ISO OSI seven-layer reference model. Specifically, bridges operate at the Media Access Control (MAC) sublayer—the bottom half of the ISO OSI Data Link layer. The IEEE bridge standards specify functions defined within the MAC sublayer. Their presence is transparent to all DTE/stations. In other words, bridges and layer 2 switches perform all of their frame-forwarding tasks without requiring modifications to the end stations. In addition to the IEEE bridge standards, there are a variety of nonstandard bridging implementations engineered to address special needs of proprietary products (see Figure 10.1).

Unlike a repeater, a token ring bridge does not forward free tokens or other MAC layer maintenance frames from one LAN to another. Therefore, all token ring LANs must provide their own free tokens, define their own token rotation time, perform Active and Standby Monitor Present functions, and assume all other ring maintenance responsibilities.

By the same "token," an Ethernet bridge does not propagate collisions occurring on any segment connected to it. Therefore, all Ethernet segments interconnected by bridges are separate Collision Domains. Most bridges and layer 2 switches adhere to the IEEE 802.1D local bridging standard and operate below the MAC service boundary. The IEEE 802.1D standard recommends a maximum of seven tiers of cascaded bridges, or bridge hops in the data path. Other IEEE 802.1 subcommittees define different bridge standards and specifications.

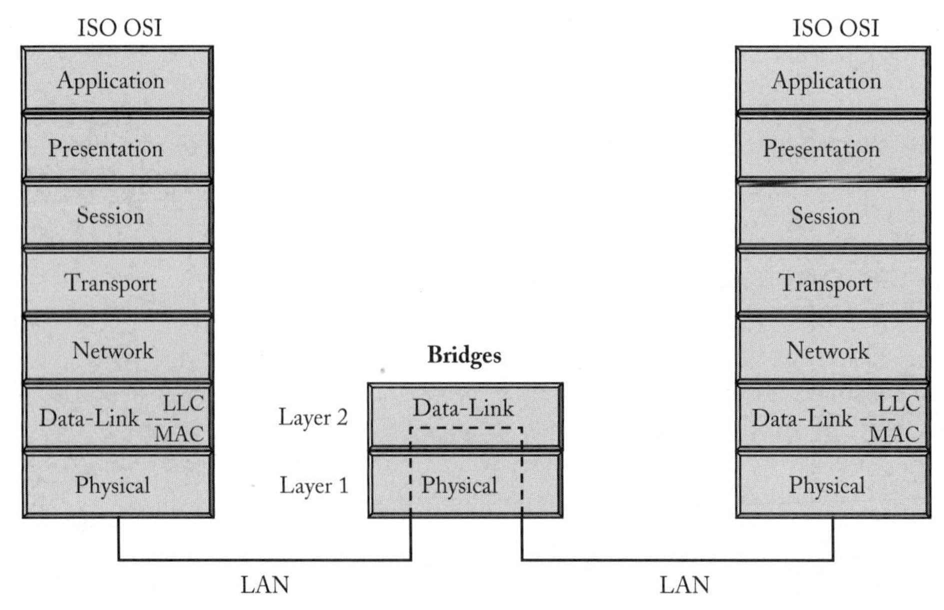

Figure 10.1 *Bridges and the OSI Model*
Bridges correspond to the functions defined in ISO OSI Data Link layer (layer 2).

The vast majority of all LAN switches are multiport bridges, and are based on the same IEEE 802.1D bridge standards. MAC bridges and layer 2 switches that adhere to the IEEE 802.1D standards are considered local, auto-learning, transparent devices. They are termed "local" devices because the 802.1D standard defines LAN-to-LAN bridging only. These devices automatically learn the MAC address of all of active DTE/stations in the network. As MAC addresses are learned, they are associated with the network interface on which that frame originated. With this information, bridges establish a database, or filtering table, which is used to make filtering and forwarding decisions on a frame-by-frame basis. The IEEE 802.1D bridge standard does specify that bridges must execute the cyclical redundancy check (CRC) algorithm against each frame and compare the results to the Frame Check Sequence (FCS) value included at the end of each frame. The purpose is to provide error detection at each bridge, with the intent to prevent the forwarding of corrupt frames. Bridges provide an effective increase in the physical extent of LANs, and increase the number of possible DTE/station attachments. Properly implemented, bridges can also improve the overall performance of large, congested networks.

IEEE 802.1D bridges forward MAC layer frames based on the MAC addresses found in the header of each frame. Frame forwarding occurs independently of the addressing contained in upper layer protocols such as IP or IPX. The standard also specifies support for the Spanning Tree Protocol (STP). All of which will be discussed later in this chapter. Layer 2 switches are high-performance devices that are easy to install and simple to use, and have become very economical; in many cases, management options are available.

Although the standard states that bridges may support the interconnection of stations attached to IEEE 802 LANs of different MAC types, such as Ethernet, token ring, and FDDI, most bridges and layer 2 switches support just one LAN technology. The media used by each bridged segment may differ, as may data transmission rates. A few vendors do sell devices that can bridge between Ethernet and token ring or FDDI. Such devices are referred to as translation bridges and require the ability to map the fields of Ethernet frames into the appropriate fields of the token ring or FDDI network frames. Otherwise, IEEE 802.1D bridges do not alter any information in the frames.

SOURCE ROUTE BRIDGES

Rather than using existing IEEE bridge standards, source route bridging (SRB) technology was developed for use with IEEE 802.5 and IBM Token Ring Network (TRN) LANs. Source route bridging is an IBM Token Ring–only technology. It is also an excellent example of confusing terminology in that it references two terms that are otherwise mutually exclusive: routing and bridging. In stark contrast to 802.1D bridging, SRB networks rely on the end DTE/stations to make the end-to-end path selection decisions over which frames are forwarded. SRB is anything but transparent to the end stations.

In contrast to IEEE 802.1D bridging, which is automatic, dynamic, and transparent, source route bridging requires a network installer to manually program each bridge with a unique bridge identifier number, and each bridge port with a unique ring identifier number. SRB supports a maximum of 4095 rings (1–4095) with up to 15 bridges per ring (1–15). Bridge ID numbers can be reused to number multiple bridges as long as no two bridges with the same bridge ID are connected to the same ring. All ring IDs must be unique (see Figure 10.2).

In SRB LANs, before a DTE/station may transmit a data frame it must issue a route discovery (RD) frame to the destination DTE/station. Each station must learn the fastest path to other stations by transmitting an RD frame addressed to its target station (server or host) at the time of session establishment. SRBs receiving such frames append their own bridge ID and the next ring ID into the route information (RI) field of the frame. The bridge ID and ring ID numbers are represented as 2-byte values with 4 bits used to define the bridge ID and 12 bits used to define the ring ID.

The standard RI field can hold a maximum of 7 SRB "hop" entries. SRBs must forward all RD frames across all rings. RD causes a flood of RD broadcasts across the entire network during session establishment. Once seven hops have been reached, the frame is considered expired and is removed from the network by the next SRB it encounters. Only those frames that successfully reach the destination station are

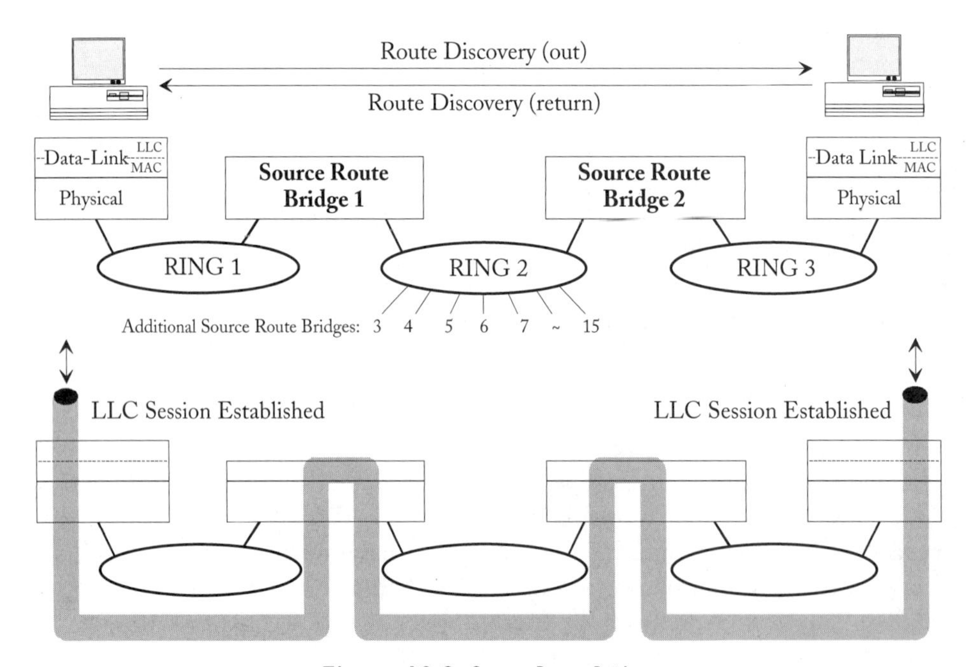

Figure 10.2 *Source Route Bridges*
Source route bridging is used only in token ring LANs.

processed and returned to the station of origin. Several proprietary solutions have been developed to extend SRB-based networks beyond the seven-hop limit.

Once the destination DTE/station (the target station) receives an RD frame it reads the contents and stores all of the RI that was written into the frames by all of the SRBs along its journey. The target station then replies by inverting the RI and retransmitting an RD reply back along the same path that the original RD request took to get there. The first frame to return to the originating (or source) station suggests the "best" path to the destination. The two stations will continue to use this path for communications between each other until the session is terminated or the communications link fails.

SRB can identify alternate backup paths between DTE/stations over a mesh of interconnected token ring local area networks, however, switching paths is rarely automatic. When applied to wide area networks, the RD broadcasts can expend huge amounts of bandwidth each time a session is established, and the response time can become intolerable.

Also, as traffic congestion levels change throughout the internetwork, SRB will not dynamically alter the established path in favor of a less congested path. In other words, SRB does not support dynamic load balancing, nor can SRB "reroute" around failed links without manual intervention (i.e., re-establishing a new session). SRB path selections remain static for the duration of the session. SRB is IBM-only, Token Ring only, and therefore represents a tiny fraction of the overall bridge market.

TRANSLATION AND ENCAPSULATION BRIDGES

Another type of bridge is the translation bridge. These devices are unique in that they can forward frames between dissimilar networks such as Ethernet, token ring, and FDDI. Some translation bridges were developed to connect Ethernet to token ring at the MAC layer, while others were developed to provide a MAC layer connection from Ethernet or token ring LANs to high-speed FDDI backbone networks. All networks interconnected by bridges, translation or otherwise, are addressed as a single logical network.

Translation bridges convert the frame structure of one LAN into the frame structure of another LAN, adding or moving some fields while deleting others. One of the problems with translation bridging is that all LANs define different field structures for their MAC frames, they also support different maximum frame lengths (see Figure 10.3).

Many translation bridges cannot break large packets down into multiple smaller packets for transmission onto a network that supports much smaller frame sizes. This function is referred to as "fragmentation" and is commonly performed by routers. The use of such limited translation bridges imposes a restriction on the frame size used to the lowest common denominator. In other words, all DTE/stations must be configured to limit their maximum frame size to one that is supported by all interconnected networks. If the translation bridge cannot fragment packets, all packets in excess of the

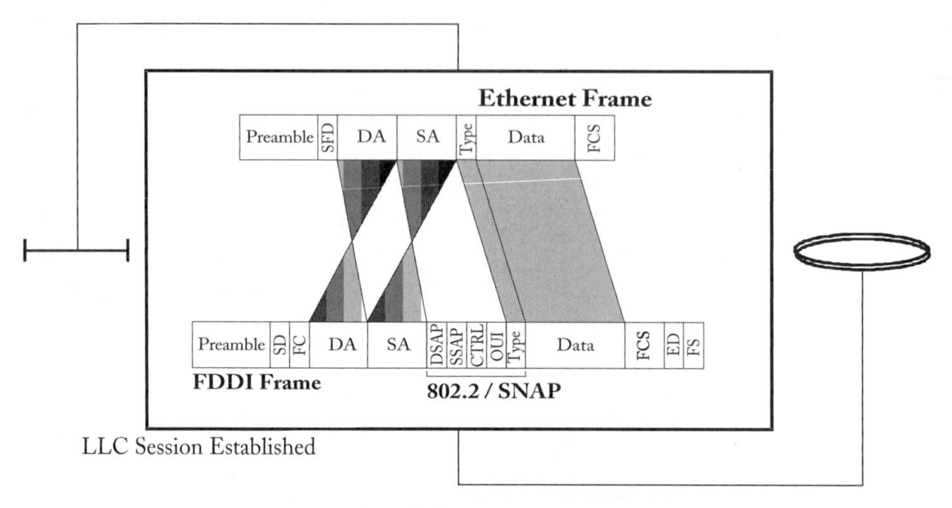

Figure 10.3 *Translation Bridges*

maximum frame size supported by the destination network will simply be dropped (discarded) at the translation bridge.

Some translation bridges designed for FDDI do perform packet fragmentation, while others use proprietary encapsulation techniques that place an entire Ethernet or token ring frame inside the data field of an FDDI frame for transport over the FDDI network. Such devices are called encapsulation bridges, and as mentioned earlier, most are proprietary. Remember that, in most cases, if something is ENcapsulated it must eventually be DEcapsulated (see Figure 10.4).

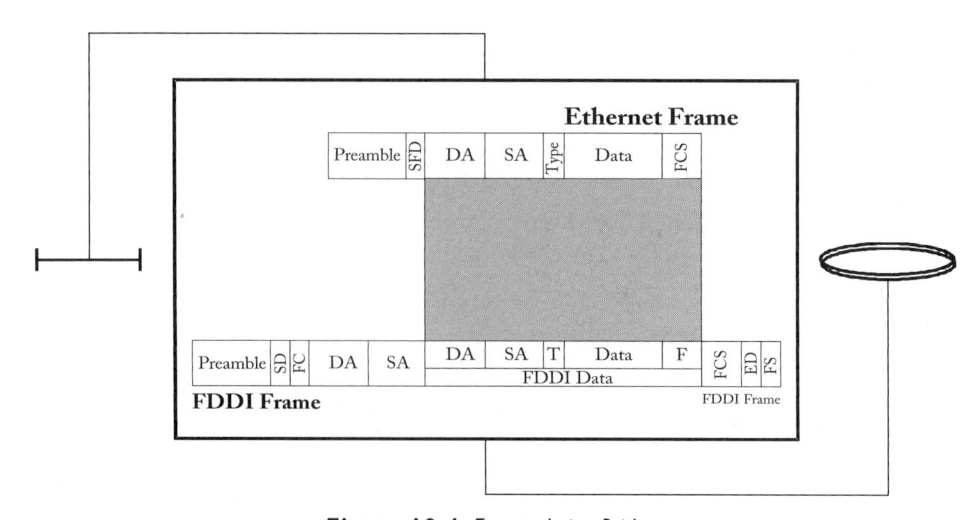

Figure 10.4 *Encapsulation Bridges*

Another problem common with both translation and encapsulation bridging has to do with the way different LANs manage MAC addresses. Some LANs read their MAC address bytes *least* significant bit first, while others read *most* significant bit first. In other words, the bits in each byte of the addresses are read either from left to right or from right to left. Obviously, reversing the bit order would result in radically different numbers. Therefore, in addition to juggling fields between different frame structures, translation bridges must also convert the MAC address format used by the source LAN into the MAC address format used by the destination LAN (and perform the reverse for each reply). Routers perform all of these functions and more while obviating such frame and address incompatibilities between the different LANs.

IEEE 802.1D TRANSPARENT BRIDGING

As mentioned earlier, IEEE 802.1D bridges are defined as local, auto-learning, transparent devices. IEEE 802.1D bridges operate at the OSI Data Link layer (layer 2) and are therefore more complex than repeaters, which operate at the OSI Physical layer (layer 1). Specifically, bridges operate at the MAC sublayer of the OSI Data Link layer. IEEE 802.1D bridges can be used to segment network traffic on large congested LANs while providing a fast, transparent link between LANs. All network segments interconnected by bridges are in the same broadcast domain. All DTE/stations within the same broadcast domain are also within the same (IP or IPX) network or (IP) subnet.

Because bridges operate at the MAC sublayer, they have the capability to examine the MAC addressing in each frame. 802.1D bridges automatically learn the location of all MAC addresses on all network segments connected to them. The 802.1D bridge program builds a filtering database of all MAC addresses based on the source address (SA) of frames transmitted by each active station attached to the network. Based on this filtering database (or forwarding table), frames will be filtered (discarded) or forwarded (regenerated and relayed) onto other LAN segments interconnected by the bridge. Bridges operate independently of higher layer protocols such as TCP/IP or IPX.

All LANs interconnected by bridges are within the same broadcast domain. This is because bridges must always forward all broadcast frames. Any MAC-layer broadcast occurring on any LAN will be propagated to all other LANs connected via bridges, or layer 2 switches. The 802.1D standard recommends traversing a maximum of seven bridge hops. This does not mean that the broadcast domain is limited to supporting just seven bridges, but rather defines the maximum number of bridge hops across the entire broadcast domain (see Figure 10.5).

Forwarding Table

In the standard (or default) configuration, 802.1D bridges operate very efficiently, imposing minimal latency on the forwarding of each frame. However, if conforming to the full store and forward processing requirement defined by the standard, each frame will experience a forwarding delay of at least the length of the frame in bits. The bridge

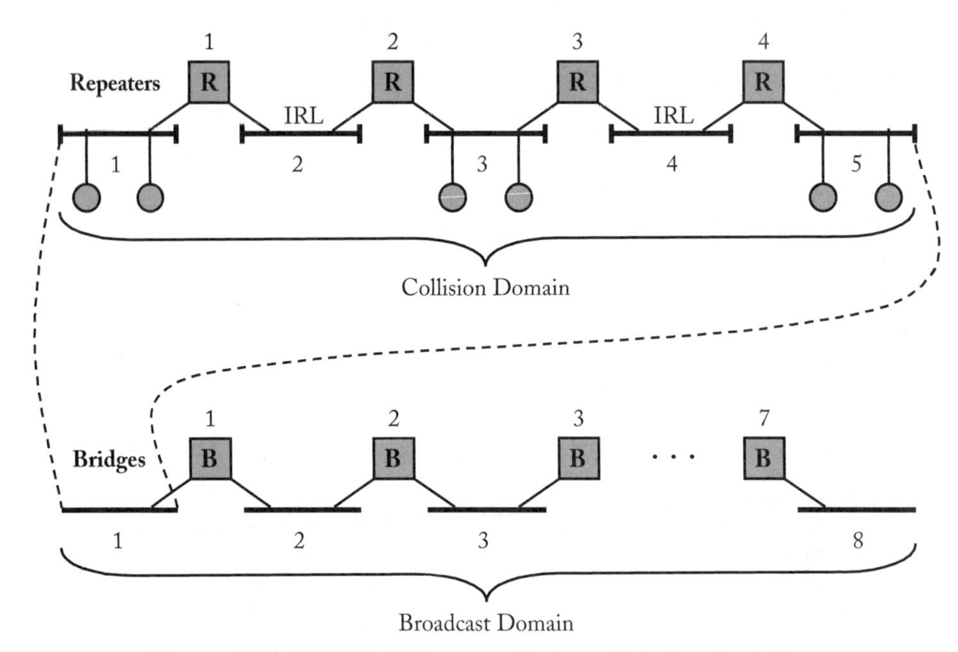

Figure 10.5 *IEEE 802.1D Transparent Bridges and the Broadcast Domain*

must also receive and hold (buffer) each transmitted frame in its entirety long enough to compare the FCS field's CRC value and examine the frame for misalignment, acceptable length, and valid addresses. If the frame is found to be corrupt, it is discarded.

For example, a 64-byte frame equals 512 bits and requires 51.2 μs to transmit over 10 Mbps Ethernet. The buffering and processing of each frame may add an additional delay, so that a 10 Mbps bridge will introduce at least 51.2 μs to forward the smallest Ethernet frame [64 bytes] and at least 1.2 ms for the largest Ethernet frame [1518 bytes]. Because of the advancements in computing technologies, internetworking hardware has benefited in higher throughput and lower latency. The performance of many internetworking products is so good today that their throughput is referred to as being equal to "wire-speed," meaning that the latency added by these components is negligible.

Token ring frame forwarding times also vary depending on the speed of each of the bridged token rings, the frame sizes supported, whether transparent IEEE 802.1D bridging or source route bridging is being used, and the performance of the bridge platform. Rather than being purpose-built devices consisting of a hardware platform engineered for the task, many early token ring (and some Ethernet) bridges were just personal computers with a pair of LAN adapters, running a LAN bridge program.

To determine which frames to forward and which to filter, IEEE 802.1D requires that all bridges (and layer 2 switches) examine all frames transmitted on any network segment attached to that bridge or switch. The MAC source address of each frame is dynamically entered into the filtering database and time-stamped to monitor the age of the entry. Each MAC address entered into the filtering database is associated with the port on which the frame was received (see Figure 10.6).

The bridging process then compares the MAC destination address in each frame to the addresses in the filtering database that are associated with that frame's port of origin. In other words, the destination address in each frame is compared to the current address entries in the database. If the destination address appears to be on the same port as that frame's source address, then they are on the same segment and the bridge does not need to forward that frame. If the destination address is *not* on the same port as the source address, then the frame must be forwarded (see Figure 10.7).

The process of forwarding frames because they are *not* found in the filtering database (associated with the port of origin) is referred to as *negative forwarding*. This protocol was used in simple two-port bridges. Today's multiport layer 2 switches require a different approach. Because they have multiple ports, switches must compare the destination address in each frame received to *all* learned addresses in the filtering database. Frames are then forwarded only to the port containing that address (but not if the destination address is on the same port as the port of origin). This is referred to as *positive forwarding*.

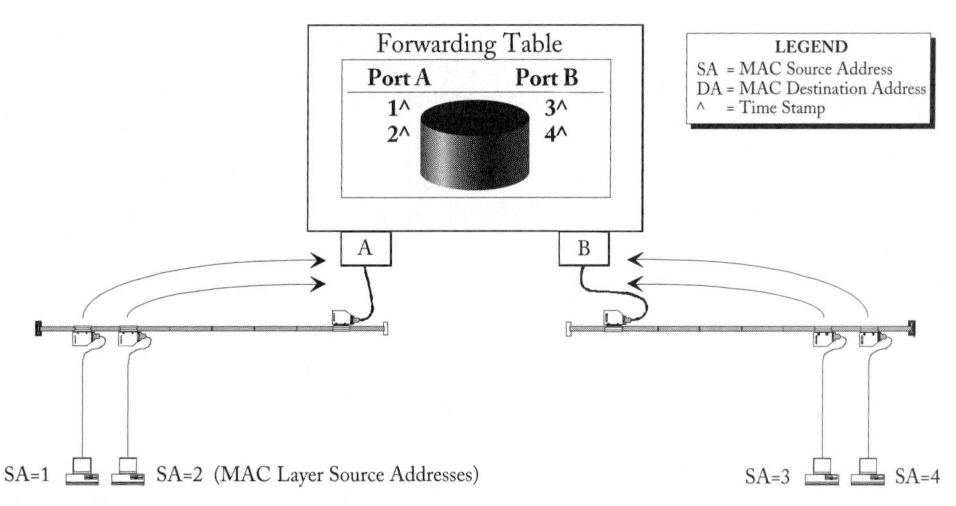

Figure 10.6 *Transparent Bridges: Forwarding Table*

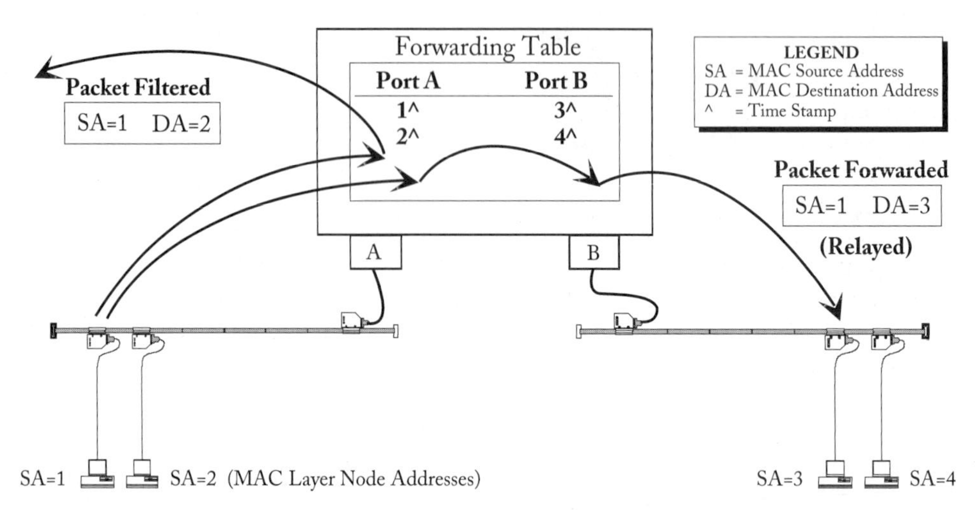

Figure 10.7 *Transparent Bridges: Forwarding Decisions*

Once a bridge determines that a frame must be forwarded, it is queued for transmission on its destination bridge port. If the frame's destination address is a broadcast address, that frame must be forwarded to all bridge ports. All bridge ports are available for forwarding as long as the Spanning Tree Protocol indicates they are in the *forwarding state*. More information on STP will be provided shortly.

The IEEE 802.1D standard specifies a recommended aging time default value of 300.0 seconds (5 minutes) for each address dynamically entered into the filtering database. However, if the bridge is manageable (via external software or a terminal interface), this value may be explicitly set to any value in the range of 10.0 to 1.0×10^6 seconds. When the time stamp of a given entry expires, that address is automatically purged from the filtering database. In this way, the database is kept lean and efficient. Reestablishing a purged address is dynamic and fast, and does not impede performance.

MAC addresses that are known by the network administrator to be permanent may be entered into each bridge manually. Manually defining MAC addresses in a filtering database is called *static forwarding*. The addresses must be manually entered in their 6-byte hexadecimal formats. These static addresses will not be relearned and redundantly entered into the filtering database, nor will they ever be purged.

Due to the dynamic nature of most networking environments, this method is not recommended. Organizations typically relocate employees and their computers quite often. Many American businesses conduct moves, adds, or changes for as many as 25% of their staff each year. Each move or change that affects the physical location of any station (with the MAC address of its LAN adapter manually entered into a bridge or switch) would require manual changes to the filtering database of all bridges and switches across the enterprise. Clearly this would be an unmanageable burden and should be avoided.

The IEEE 802.1D MAC bridge standard supports all IEEE and ANSI standard LANs, including 802.3/Ethernet, 802.5 token ring, and ANSI X3T12 FDDI. Most bridges support the same IEEE network access method on each of their ports—i.e., either IEEE 802.3/Ethernet, IEEE 802.5/token ring, or ANSI X3T12. There are few, if any, bridges available for non-standard LANs.

SWITCHING METHODS

Switch designers have had differing opinions about which architecture provides the most efficient mechanism for forwarding LAN frames. Since the development of switches, some manufacturers have used a cross-point switch matrix, while others have implemented a shared memory approach. One solution found to provide the best performance and economies of scale is achieved with a silicon-based switched mesh fabric using Application-Specific Integrated Circuit (ASIC) technology. As the name implies, ASICs are silicon chips custom designed to perform specific tasks. Today's switch products rely on arrays of PHY chips that provide the interface between the Physical layer protocols (100Base-TX) and media (UTP). The PHY chips interface to one or more MAC chips that perform the (802.1D) bridging tasks.

The terminology in this market is as varied as the products. The term "switch" can have many meanings depending on one's personal education and experience, and the current networking environment in question. For example:

> A **Circuit Switch** usually refers to a Private Branch eXchange (PBX) used in telephony systems. As a call is placed, telephone switches dynamically establish (or "nail up") a circuit between stations. This circuit is maintained for the duration of the call, or session, whether any data is transmitted or not. When the call is concluded the switches terminate (or "tear down") the circuit, thereby releasing those resources to support other calls.

> A **Frame Switch** is a LAN-based device that typically provides the functionality of a multiport MAC layer bridge. These devices are most often referred to as layer 2 switches and are the most common form of LAN switch. They dynamically forward MAC layer frames between stations connected via any port on the device. Layer 2 switching occurs on a frame-by-frame basis rather than per circuit or session. Most layer 2 switches implement IEEE 802.1D standard MAC layer bridging, Cut-Through or Fast Forward, or Cut-Through with Fragment-Free or Cut-Through with Error-Free forwarding. Cut-Through (Fast Forward) reduces the latency by forgoing the CRC check and allows the switch to begin forwarding a frame immediately. Error-Free (Fragment-Free) operates in a similar manner to Cut-Through (Fast Forward), except that a minimum of 512 bits of each frame is buffered before being forwarded. In a correctly built network, this eliminates the possibility of forwarding a frame fragment.

Several LAN switches function as multiport layer 3 routers and are referred to as layer 3 switches. As previously mentioned, several vendors are also promoting products they call layer 4 (or even layer 5) switches.

Cell Switches are based on Asynchronous Transfer Mode (ATM). ATM switches establish permanent virtual circuits (PVC) or switched virtual circuits (SVC) over which fixed-length 53-byte cells are forwarded. Whereas all other network technologies support variable length frames, ATM uses fixed length packets, called cells. All ATM cells have a 5-byte header, a 48-byte payload, and no trailer.

ATM switches are typically layer 2 devices. ATM switches forward fixed-length cells much like LAN switches forward variable-length frames; however, ATM requires virtual circuits and provides several different service levels, defined in the ATM Adaptation Layer (AAL). While ATM can be used for enterprise, campus, local area and wide area network applications, most organizations find the ATM infrastructure to be too complex and expensive to implement and maintain. ATM is enjoying some of its greatest success at the core of major carriers' networks.

LAYER 2 LAN SWITCHES

Today's Ethernet switch industry offers a plethora of high-performance products, from large, high-capacity and highly complex enterprise switches to small, simple, and economical switches engineered for the small office/home office (SOHO). From OSI layer 1 through layer 5, there is a switch for every niche. Layer 2 switches are essentially very fast and efficient multiport bridges and have almost completely replaced simple repeater/hubs as the preferred method of networking an organization's computers. Switching technology has also been leveraged to improve the power and performance of routers, which are referred to as layer 3 switches. Layer 3 switches are more complicated and somewhat more expensive than their layer 2 siblings. In summary, layer 2 switches are multiport bridges and most layer 3 switches multiport routers.

The primary purpose for implementing layer 3 routing in a switch is to reduce latency and improve packet throughput. Most traditional routers include an array of packet filtering features, network interface options, security and management functions, and advanced internetworking capabilities that most layer 3 switches will lack. The packet filtering, security, and management functions in routers tend to add delays to packet forwarding, therefore increasing latency. Since the objective of using layer 3 switches is to reduce latency, enabling such functionality would be counterproductive. Because of this, both layer 2 and layer 3 switches are usually deployed within an organization's private internal network (or intranet), whereas traditional routers are deployed at the boundary to the Internet or to another intranet. The switches used within the intranet typically require only fast or high-speed LAN interfaces and lit-

tle or none of the packet filtering or security functions, whereas the boundary routers typically must connect to some sort of WAN, usually provide advanced packet filtering, additional security features, and some degree of management.

As with computing horsepower, and memory and disk drive speed and capacity, the switch industry's prices have dropped rapidly as features and performance increased dramatically. Prolific competition and the rapid expansion of the Ethernet switch market continue to drive prices down and performance up. The earliest switches neglected such features as network monitoring, management, and protocol analysis support, but LAN switch developments continue to improve the performance and add new features with each new generation of product released.

In addition to standard network management capabilities, many switch vendors are now including a "trace port" that allows a network administrator to promiscuously monitor and capture traffic on any or all ports of the device. Most devices now include Web-based network management to provide access to advanced protocol filters, prioritization and Quality of Service (QoS) mechanisms, VLAN configuration, network performance monitoring, and troubleshooting capabilities.

Still, it is important to realize that all switches are not created equal; some vendors adhere to standards and some blaze their own trails. All switches are not engineered to satisfy the same requirements; some products target the SOHO market while others are engineered for the enterprise. All switches do not use the same frame-forwarding protocol; while some still support full 802.1D store and forward, others implement cut-through to reduce latency. All switches do not operate at the same layer of the OSI model; some operate at layer 2 (bridges), or layer 3 (routers), while others offer special functionality relying on session-specific and/or application-specific information contained at layers 4 and 5.

Some layer 2 LAN switches may implement all or just a portion of the IEEE 802.1D bridge specifications. For example, a layer 2 switch may be designed to forward frames as soon as the destination MAC address is ascertained. Since the destination MAC address is located in the first 6 bytes of each Ethernet frame header, a switch can read the address, compare it to the forwarding table, and make its forwarding decision well before the entire frame has been received by the switch. If the switches are not required to execute the CRC algorithm prior to forwarding, they do not have to buffer the entire frame. Such a switch can achieve higher throughput and lower latency by starting to transmit each frame onto its destination segment as soon as the address look-up has been completed. This latency-reducing technique is referred to as cut-through or fast forward.

Since layer 2 switches forward MAC frames based on MAC addresses, they can segment traffic like any MAC bridge, the purpose being to keep local traffic local and forward traffic only to segments where it belongs. This minimizes wasted network bandwidth and helps to improve overall network performance. However, if the switches do not perform error detection on each frame before forwarding them onto the frame's destination network segment, corrupt or misaligned frames can be forwarded.

As mentioned previously, the minimum Ethernet frame size is 512 bits. Frames smaller than 512 bits are called fragments, or in some cases runts, and are interpreted as collisions by all standard Ethernet devices. It is possible that the header of a frame fragment may contain an intact destination MAC address. A cut-through switch may successfully read the destination address, compare it to the forwarding table, make its forwarding decision, and begin transmitting onto the destination segment, never detecting that the frame is corrupt. The result is that some of these switches do not completely segment Ethernet Collision Domains. Fortunately, most modern Ethernet LANs exhibit extremely low error rates, thereby minimizing this potential problem. And in full-duplex environments, collisions are eliminated.

LAN switches can forward multiple frames concurrently, resulting in data streams that generate an aggregate throughput far in excess of the capacity of any single network interface. Remember that switches also permit full-duplex transmissions that can double the effective throughput demands on the switch. Because of this, all switches, especially high-end switches with higher port densities, must support an aggregate bandwidth capable of forwarding frames at twice the data rate of all interfaces combined. In other words, a 12-port 100 Mbps switch should be engineered to support full-duplex transmissions on all 12 interfaces simultaneously, or 2.4 Gbps (12 ports × 200 Mbps).

Many high-end LAN switches also support high-speed up-link connections such as FDDI, ATM, multiple aggregated Fast Ethernet segments, or Gigabit Ethernet. Of all of these options, Gigabit Ethernet has become the preferred high-speed backbone for interconnecting LAN switches. In addition, some LAN switches also support Virtual LAN (VLAN) capabilities and one or more protocol prioritization schemes. VLAN capabilities usually include support for IEEE 802.1Q. Protocol prioritization methods may include IEEE 802.1p service class switching, Multiprotocol Label Switching (MPLS), or Differentiated Services (DiffServe).

IEEE 802.1D SPANNING TREE PROTOCOL

Loops in Ethernet networking can be disastrous, causing erroneous collisions and MAC layer broadcast storms. The intelligent application of loops as backup paths can provide link redundancy (fault tolerance). The STP was developed to prevent loops and to allow for redundant bridged links to provide alternative paths around failed segments of the network.

STP is a bridge-to-bridge link management protocol implemented in all IEEE 802.1D–compliant devices. The protocol implements a distributed algorithm used to determine the topology of the spanning tree. When multiple paths exist between two or more bridges, STP will choose one bridged path as the primary path and place it in the forwarding state, while placing the other redundantly bridged paths into the blocking state, thereby interrupting the loops. There are five states defined by STP. Each bridge assigns a state to each of its ports. These states determine when MAC addresses may be learned, when it may forward frames, or when it must block loops. The five states of the Spanning Tree Protocol are:

Disabled	Entered when the network manager explicitly disables a bridge port.
Forwarding	The normal state where addresses are learned and frames are forwarded.
Blocking	Initiated by STP to interrupt loops. Addresses are not learned, frames are not forwarded.
Listening	Entered from Blocking state. Initiated by STP to begin the transition from Blocking to Forwarding.
Learning	Entered from Listening state. Addresses are learned, but frames are not yet forwarded.

The Listening and Learning states are intermediate states that a port goes through as it transitions from Blocking to Forwarding. Their purpose is to prevent loops from being created during network reconfiguration. STP prefers a temporary partitioning of a network segment to a loop.

STP interrupts redundant paths by placing selected bridge ports into the blocking state. A bridge port in the blocking state is not actively contributing to network capacity or otherwise handling any traffic flow. It is waiting for a fault to occur. Its purpose is to provide an automatic hot-swap of network paths, without manual intervention. STP does perform this function, just not as quickly as one might expect. By default, most STP implementations will require about 45–90 seconds to determine that a fault has occurred and correct the situation. During this period, most network communication protocols will time out and report a network failure to the user (i.e., Abort, Retry, Fail). Two or more retries may be required before STP resolves the problem, by which time most users will have rebooted their computer and called the help desk twice.

The STP process starts when the bridge or switch is powered up. Each STP device begins by assuming that it is the root of the spanning tree and periodically transmits a "hello" message on all ports announcing its presence. The "hello" message is actually a Bridge Protocol Data Unit (BPDU). Each bridge continues to transmit this message until it receives a message specifying another bridge with a lower bridge Port Identifier value. By default, the bridge with the lowest MAC address becomes the Root Bridge (see Figure 10.8).

If a bridge receives BPDU messages from other bridges contending for root status, it responds with a message indicating its choice of root to correct the less suitable

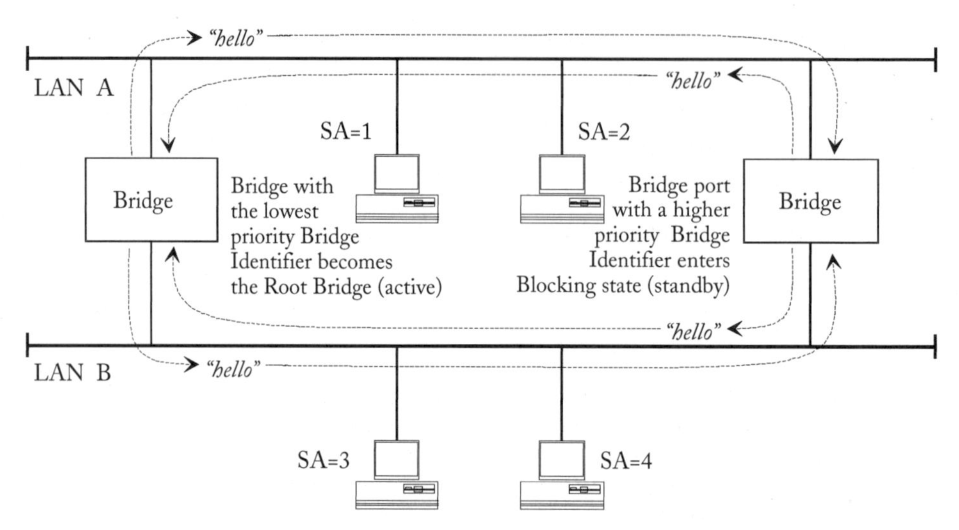

Figure 10.8 *IEEE 802.1D Transparent Bridges: Spanning Tree Protocol*

candidates. This process usually stabilizes relatively quickly resulting in a single network-wide choice of root (see Figure 10.9). If it does not stabilize, or converge, the bridged internetwork will continually reset and reconfigure, resulting in loss of access and conflicting STP tables. The phenomenon can be caused by the implementation of incompatible versions of STP in a common network.

There are several versions of the STP currently in use, including the early DEC version as well as the IBM and IEEE standard versions. For example, Cisco routers and switches support all three versions, but consistency cannot be presumed. Cisco routers default to the DEC version of STP, while the Cisco switches default to the IEEE version of STP. These different versions are not compatible with each other. Pick one (IEEE) and use it everywhere.

All bridge and switch ports must have unique MAC addresses. STP will also assign each port a unique Port Identifier. In addition to the MAC address and Port Identifiers, there is a Path Cost associated with each bridged connection. If the bridge supports software management, the network manager can manually define the Path Cost value. Usually, the higher the speed of the network or link, the lower the Path Cost value. Each bridge attempts to determine which path has the lowest Path Cost to the root. This will become the *designated path*.

The IEEE STP specifications recommend that default values of the Path Cost parameter for each bridge port be based on the following formula:

Path Cost = 1000/Attached LAN speed in Mbps
(Which gives a Path Cost of 10 for a 100 Mbps LAN)

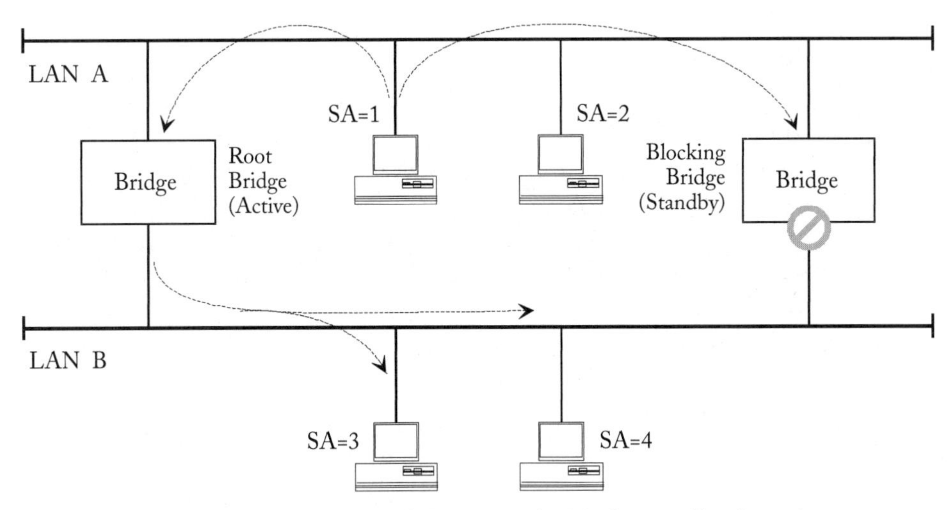

Figure 10.9 *IEEE 802.1D Transparent Bridges: Spanning Tree Protocol*

If a multiport layer 2 switch determines that two or more of its networks provide a path to the root, it stops forwarding on all but the network port with the lowest Path Cost and places the remaining ports in the Blocking state. If several switch ports have the same Path Cost, the port with the highest priority Port Identifier remains in the Forwarding state. The network manager can also explicitly define port priorities.

Once a port is placed in the Blocking state, it no longer transmits or receives any frames. The state of the spanning tree is constantly monitored for changes. If one of the network segments composing the tree becomes inaccessible, STP will automatically reconfigure the topology of the tree to reestablish a fully connected extended network. To accomplish this, ports previously placed in a standby state are enlisted into active service. Thus, the STP not only resolves and interrupts loops, but also allows redundant paths to be configured and maintained as hot standbys.

SPANNING TREE PROTOCOL PROBLEMS

Layer 2 switches are often deployed to directly support user PCs, workstations, and servers in place of hubs. Since most layer 2 switches adhere to the 802.1D bridge standard, they also support STP. This can cause certain problems for switch-attached computers. One of the problems depends on whether those computers have been assigned static network address information (such as IP addressing) or whether they obtain their IP address information via a Bootstrap Protocol (BootP) or Dynamic Host Configuration Protocol (DHCP) server.

When a computer is powered up it performs a power-on self-test (POST). This is the initial step in the normal bootstrap process, sometimes called "boot-up" or Initial

Program Load (IPL). As the operating system loads, it begins to initialize peripherals such as the network interface card (NIC). Once the NIC is initialized it asserts a Link Pulse (10Base-T) or Fast Link Pulse (100Base-TX). This Fast Link Pulse is incorporated into all 100Base-TX network interfaces used in today's computer NICs, switches, and routers. Once a switch senses the Fast Link Pulse on an interface it responds in kind to auto-negotiate for the highest speed and full or half duplex. Once the link speed and duplex have been determined, Spanning Tree Protocol on the switch then begins to transition from the Blocking state, to Listening and Learning, and finally to the Forwarding state.

Hopefully, by the time the computer begins to load the operating system the NIC has initialized and the link speed and duplex have already been negotiated. Early on in the bootstrap process, the operating system will load its network protocols. Assuming that TCP/IP is one of the protocols and static IP addressing has not been assigned, the IP protocol stack will broadcast a DHCP request onto the network to request an IP address and related information from a DHCP (or BootP) server. The problem arises if STP has not completed its transition into the Forwarding state by the time the DHCP request is transmitted. As previously stated, the STP state transition delay can take as long as 45–90 seconds. Unless the switch port has transitioned into the Forwarding state, the DHCP request will not be forwarded. The computer's DHCP request will eventually time out, and report that no DHCP server could be found. Since the computer will fail to obtain an IP address via DHCP, it will therefore lack the IP addressing information needed to communicate over the network.

STP on switches is also a nuisance for network technicians who find themselves waiting for what seems like an eternity for STP to start forwarding before they can establish and test dubious network connections. To overcome these hurdles, some switches such as the Cisco Catalyst series provide an option called *PortFast* that circumvents the tedious STP state transitions. PortFast must be configured on each interface where circumventing STP is desired. With PortFast, STP is still enabled, but it transitions directly from Blocking to Forwarding, thereby skipping the Listening and Learning states. The risk in doing this is that there exists the potential of creating a loop between multiple switch ports, or between multiple switches. If a loop is created between bridges/switches, the resulting MAC layer broadcast storm will quickly bring the entire broadcast domain to a halt. However, with PortFast, STP will quickly detect the loop and place one of the offending switch ports into the Blocking state, thereby breaking the loop. Whereas PortFast provides the benefits of STP without the annoying transition delays it normally imposes, another alternative is to disable STP entirely. This can usually be performed on a per-port basis, or STP can be completely disabled on the entire switch.

OTHER SWITCHING PROBLEMS

Switches are excellent devices for managing network capacity and traffic flow; however, they do create some difficulties for network technicians. Two commonly used tools of the trade are network monitors and protocol analyzers, both of which must be tapped into the network data stream to monitor and analyze network traffic. This is easy to do in a repeater/hub-based network because hubs propagate all data streams to every port. Just plug the monitor/analyzer into any port on a hub and it will see all traffic generated by all stations in the Collision Domain. Since switches bound the Collision Domains, and forward frames only to those ports corresponding to the destination address in each frame, they prevent the monitor/analyzer from seeing all network traffic. In fact, the only traffic that would be forwarded to any switch port would be frames destined to MAC addresses located on that port, and broadcasts. Since no station on the network is likely to try to communicate with the monitor/analyzer, only broadcast traffic would be seen, which negates most of the value of the monitor/analyzer.

To overcome this problem, some switch manufacturers have added a special trace port or monitor port, which can be programmed to replicate traffic from any single switch port or group of ports. The network monitor/analyzer can be plugged into this special port and the desired traffic flows can be copied to the monitor/analyzer. This solution is less than perfect, but it is better than nothing. Another imperfect solution is to insert a hub between the stations in question and attach the monitor/analyzer to the hub. Of course, this configuration can support only half-duplex connections.

CHAPTER SUMMARY

Bridging comes in many forms, including encapsulation, translation, source route, and IEEE 802.1D transparent with Spanning Tree Protocol. Of all of these, 801.2D is the most common. The term "bridge" has fallen from favor, probably due to marketing pressure and the stigma the term carries as an archaic device. Today's bridges are known as layer 2 switches. Layer 2 switches still adhere to most of the 802.1D specifications, but may cut a few corners to improve performance. Spanning Tree Protocol may have outlived its usefulness, but some manufacturers have added a few improvements. Switches, especially 10/100/1000 Mbps Ethernet switches, are preferred over LAN hubs. In fact, some applications may require a switch-based LAN infrastructure to satisfy performance objectives.

REVIEW QUESTIONS

1. Bridges operate at what layer of the OSI model?
2. Which LANs support source route bridging?
3. What is the purpose of route discovery?
4. Which LANs support 802.1D transparent bridging?
5. What is the purpose of Spanning Tree Protocol?
6. What is the difference between bridges and layer 2 switches?
7. Which do bridges forward: packets, frames, or cells?
8. Define the following acronyms:

SRB	STP
RD	BPDU
ATM	ASIC
PVC	SOHO
SVC	VLAN

9. Identify two potential problems with STP.
10. Which network relies on permanent virtual circuits?

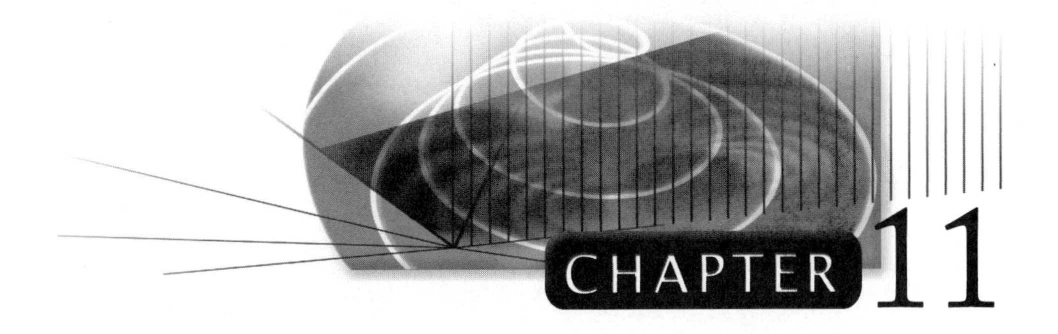

Internetworking Devices: Routers and Layer 3 Switches

OBJECTIVES

After completing this chapter, you should be able to:

- Explain the relationship of routers and layer 3 switches to the OSI model.
- Identify the differences between routers and bridges.
- Explain the difference between dynamic and static routing.
- Identify the various interior and exterior gateway protocols.
- Define the difference between routable and non-routable protocols.
- Identify the differences between RIP and OSPF.
- Identify several popular operating systems that provide routing functions.
- Explain the concept of OSPF areas and route summaries.
- Identify the problems with RIP-based routing.
- Explain the basics of autonomous systems and routing policies.

INTRODUCTION

This chapter will address another class of network interconnection components: routers and layer 3 switches. These devices extend networks well beyond the topologies supported by bridged networks. Whereas bridges can be used to network different Ethernet Collision Domains, routers are used to internetwork different broadcast domains. Although there are many routable network protocols, the emphasis in this text will be on TCP/IP. Again, as a basis for comparison, all devices will be defined according to their relationship to the ISO OSI seven-layer reference model. This will help to provide an understanding of the function of each device and the relationship to other devices.

AN OVERVIEW OF ROUTERS

Router functions are defined by layer 3 of the ISO OSI seven-layer reference model—specifically, the Network layer. Because routers operate at layer 3 they are therefore more complex than bridges, which operate at layer 2 (see Figure 11.1).

Routers rely on network protocol–specific addressing and routing information to determine the location of each device or host on the internetwork. Routers are essentially computers; they have one or more processors, memory, BIOS, and operating system, and run a couple of programs that perform the routing tasks.

Network protocols include the Internet Protocol (IP), Novell IPX, Digital DECnet, Apple Computer Corp. AppleTalk, and several others. Each of these network protocols is considered *routable* because they provide Network layer addressing used by routers to make packet-forwarding decisions. Other protocols such as Microsoft NetBEUI, IBM SNA, and Digital LAT are considered non-routable. Many routers support IP-only, and some support multiple protocols in addition to IP. In any case, all routers today support IP, are network protocol–dependent, and will only forward those packets generated by protocols they are programmed to understand.

Complex internetworks involve both a physical design and a logical design. The physical design of a network is implemented via the mechanics of the horizontal and backbone cable plant, including the hubs and switches that connect everything. The

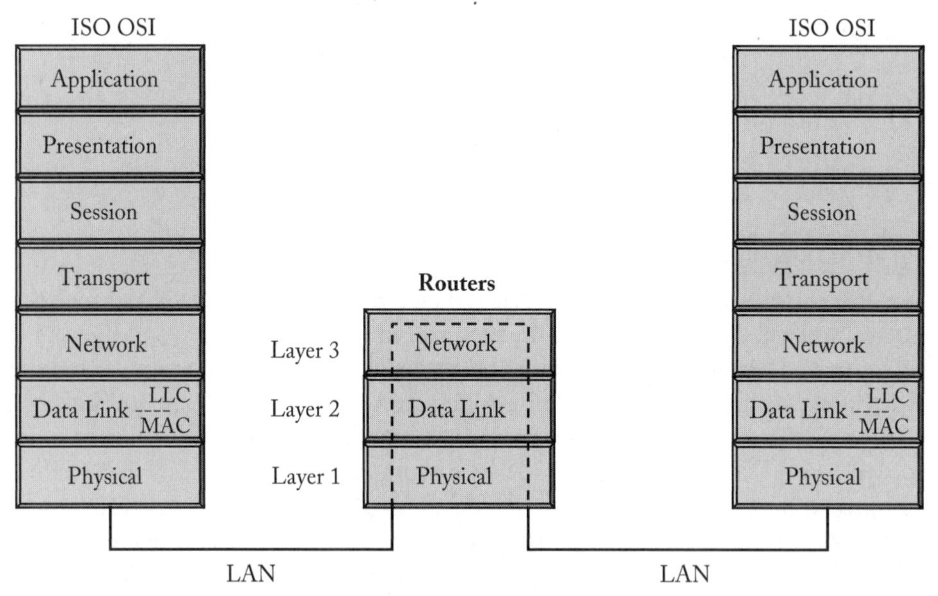

Figure 11.1 *Routers and the OSI Model*
Routers correspond to the functions defined in ISO OSI Network layer (layer 3).

logical design is implemented via routers using a variety of addressing, subnetting, and routing schemes. The flexibility and extensibility of the logical design of an internetwork is limited by the protocols used. It is also important to note that all routers do not support all protocols. For example, most SOHO routers offer only a very limited menu of routing protocols and options.

Routers make their packet-forwarding decisions based on information contained in a routing table. The information in the routing table includes the address of all networks and an associated metric, or path cost value. If this information is entered into each router manually, it is referred to as *static routing*. If each router dynamically learns this information it is referred to as *dynamic routing*. Whereas static routing depends on input by a network administrator, dynamic routing relies on one of several *routing update protocols*, such as RIP or OSPF.

Routing update protocols determine the best path to each network across the internetwork. Depending on the routing update protocol employed, the term *best path* may mean least number of router hops between stations, the shortest or fastest path, the most economical path, or a combination of these and other metrics. For some routing update protocols, the selection processes and their variables are inaccessible to the user, while other routing update protocols allow path selection parameters to be manually tuned.

The only way to move a packet (such as IP) from one network or subnet to a different network or subnet is via a router. Some small internetworks and most large, complex internetworks require the use of routers. This means one or multiple sites that require one or multiple LAN interconnections supporting one or multiple routable network protocols. The LAN and WAN technologies used may be diverse or consistent across the internetwork. Proactive network management via some Simple Network Management Protocol (SNMP) network management console is typically a requirement for larger systems. Routers are also desirable if multiple remote sites are to be interconnected in a fault-tolerant mesh configuration with a combination of leased private lines, ISDN, SONET, frame relay, or other WAN services.

Routers provide a finer granularity of control over network traffic flow and much more powerful and flexible management of enterprise internetworks than do bridges or layer 2 switches. Surprisingly, due to recent technological advancements, routers in the form of layer 3 switches are frequently as fast and efficient as layer 2 switches. Routers are also more flexible, since they forward information based on the higher-level network layer protocols and network addressing schemes.

For some systems, however, routers may be entirely inappropriate or not work at all. For example, the Digital LAT protocol, IBM's implementation of NetBIOS, and Microsoft's NetBEUI contain no Network layer address information and are therefore non-routable protocols. To accommodate such protocols, many vendors offer bridging or encapsulation options for their routers. Without such options, a router will never forward frames generated by these protocols. Fortunately, most enterprise networks do not rely on these protocols.

Routers can also provide concurrent communications between several different local area network technologies such as Ethernet, Fast Ethernet, Gigabit Ethernet, most flavors of token ring, FDDI/CDDI, and ATM. However, routers only provide communications between end stations (or hosts) using the same network protocols such as TCP/IP, IPX, DECnet, or AppleTalk. For example, an IPX host cannot communicate directly with a TCP/IP host. Routers are not protocol converters or translators; that functionality is performed by protocal gateways.

As with bridges, the processing of each packet creates a forwarding delay or additional latency. Many of the newest routers are very efficient and can outperform some bridges. In recent years, layer 2 LAN switches have been promoted as high-performance, low-latency replacements for routers, but layer 2 switches cannot forward packets between different networks or IP subnets. Most layer 2 switches are merely bridges, and all networks or segments interconnected by bridges are in the same broadcast domain. All stations or hosts within the same broadcast domain are also occupying the same subnet. Routers forward packets between networks or subnets and they bound broadcast domains.

Unlike IEEE 802.1D bridging, there are no IEEE rules regarding routers, router hops, or multiple tiers of routers. Whereas the 802.1D standard recommends a maximum of seven bridge hops between all stations, routers have no such fixed limit. Routers are limited only by the routing update protocols used, can internetwork multiple broadcast domains, and vastly extend an internetwork to support very large and complex internetwork topologies (see Figure 11.2).

Most internetworking protocols are proprietary, the support and design assistance for which is available only from the vendor. However, all routers support the Transmission Control Protocol/Internet Protocol (TCP/IP). TCP/IP represents a robust internetworking protocol suite that has evolved over the years to become the de facto standard in internetworking worldwide. Considering that TCP/IP is not technically a formal standard, it is probably the best documented, and certainly the most popular and most written-about protocol, around. Information about the entire TCP/IP suite of protocols is available free of charge to the public and can be found on the Internet as a series of more than 3000 Requests for Comment (RFC). RFC status can be confusing because these documents define draft proposals as well as ratified internetworking specifications. The protocols and policies defining TCP/IP are created and maintained by the Internet Engineering Task Force (IETF), which is part of the Internet Architecture Board (IAB). However, ANSI has been considering a proposal that would allow the organization to formalize TCP/IP as a standard.

A router is required any time packets must be forwarded from one subnet to another. Each port, or interface, on a given router represents a unique network or subnet. No

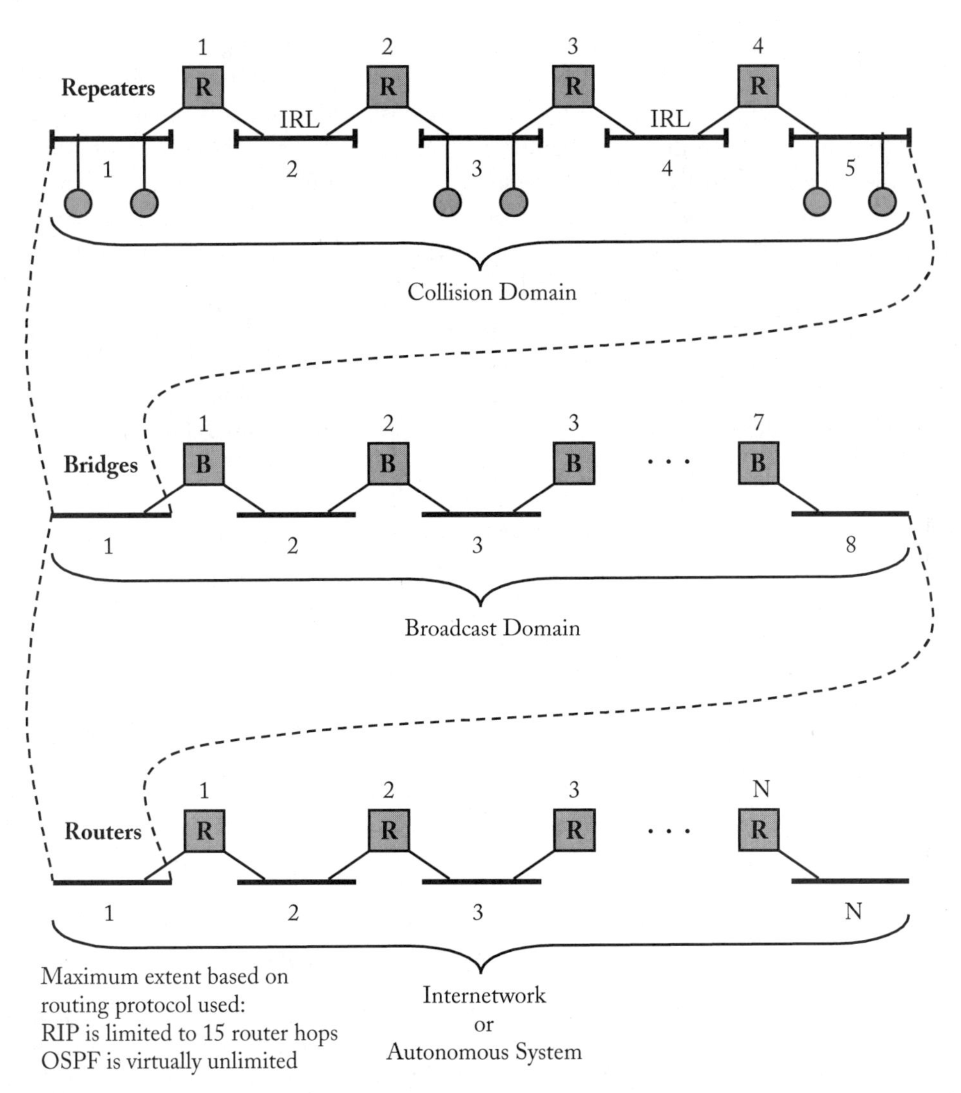

Figure 11.2 *Routers and Autonomous Systems*

two interfaces on a given router may connect to the same IP subnet (there are special exceptions). A single router port may support multiple IP subnetworks; this is called *secondary IP addressing*. All hosts (computers, printers, and other IP devices) on a given LAN segment must be configured with an IP address that is within the subnet(s) defined by the router interface to which that segment is attached.

In IP parlance, every network-attached device with an IP address is considered a "host," including mini-computers, mid-range computers, mainframe FEPs and Controllers, Unix systems, PCs, Macs, LAN-attached printers, wireless IP devices such as Personal

Digital Assistants (PDAs) and IP-enabled cell phones, all router ports, and all SNMP manageable devices on the network. A good rule of thumb is to try to limit the number of hosts per subnet to about 120. This provides for a reasonably sized workgroup of users and shared resources and can be accommodated with a 7-bit IP host address field (which can support 126 host addresses). When subnets get much larger than this they tend to become unwieldy, and difficult to manage and troubleshoot; much smaller and they are easily outgrown. Keep in mind this is just a recommendation, not a law.

ROUTING UPDATE PROTOCOLS

A very important, somewhat complicated issue is that of the various routing update protocols that perform route table maintenance and path selection. These all-important protocols operate invisibly in the background to dynamically maintain a table of all available networks and the various paths to reach them. These protocols, in effect, are inter-router communication protocols that allow the routers to exchange information over the same network links that carry user data. Without a dynamic routing protocol, static routes must be manually entered into each router. Static routes are adequate in small, simple internetwork configurations; however, dynamic routing becomes advantageous as networks become larger and meshed. There are two basic classes of dynamic routing protocols: Interior Gateway Protocols and Exterior Gateway Protocols.

Several different Interior Gateway Protocol (IGP) options exist just to support TCP/IP, such as RIP, RIP-II, IGRP, E-IGRP, and OSPF. The most popular options include the IETF's Routing Information Protocol (RIP), Open Shortest Path First (OSPF), and the use of static routing. Many large Cisco shops use Cisco's own Interior Gateway Routing Protocol (IGRP), or the newer Enhanced Interior Gateway Routing Protocol (EIGRP). Another option that can also support TCP/IP is the ISO's Intermediate System to Intermediate System (IS-IS), but it never gained wide acceptance.

Table 11.1 compares four of the major interior gateway routing protocols, the type of technology used, the number of router hops supported, and the organization or company responsible for their development.

TABLE 11.1 INTERIOR GATEWAY PROTOCOL COMPARISON

Protocol	*RIP*	*OSPF*	*IGRP*	*EIGRP*
Type	Distance Vector	Link State	Distance Vector	Balanced Hybrid
Router Hops	15	Unlimited	15	Unlimited
Developer	IETF	IETF	Cisco	Cisco

Different routing update protocols collect and propagate different kinds of information about the networks for which they are responsible. Some protocols collect network addresses and calculate a metric, or measurement, of how many router hops each network is from each router. These are called Distance Vector protocols. Other protocols collect the network address and several more complex details that describe the fastest or most efficient path to each network, and maintain information on the status of each path or link. These are called Link State protocols. While some of these protocols are responsible for maintaining routing information for all networks within an entire internetwork, others are only responsible for maintaining information about a subset of routers in a given portion of an internetwork.

These protocols actually operate over their respective network protocols to provide inter-router communications. For example, RIP uses the User Datagram Protocol over IP (UDP/IP) as a vehicle to allow IP routers to broadcast their route tables to other IP routers. Routers must communicate with each other to share information about the current status of each of their network links. This information in essence creates a "map" or directory of reachable networks within the internetwork topology. Within the class of Interior Gateway Protocols there are several variations: Distance Vector (RIP & IGRP), Link State (OSPF & IS-IS), and balanced hybrid (E-IGRP).

There are also several different Exterior Gateway Routing (EGP) protocols in use, emerging, or retired, such as GGP, EGP, BGP, and IDRP. The most popular is Border Gateway Protocol version 4 (BGP4), and again, the use of static routing. The older EGP has been displaced by BGP, the Gateway-to-Gateway Protocol (GGP) is obsolete, and the ISO's Inter-Domain Routing Protocol (IDRP) is irrelevant.

Exterior Gateway Protocols are used to provide routing information between disparate internetworks, or different autonomous systems (AS). All routers in a common internetwork, sharing a common routing policy, are within the same autonomous system. For example, the connection between a company's internetwork and an Internet Service Provider (ISP) requires an EGP, or static routing must be defined manually. The private internetwork represents one autonomous system with its own routing update protocols and policies, and the ISP represents another autonomous system with different routing update protocols and policies. The private company and the ISP must agree only on the EGP, its policies, and addressing used over the common link between the two organizations (see Figure 11.3).

ROUTING INFORMATION PROTOCOL

RIP is a Distance Vector protocol that consists of a simple hop-count metric. It does not take into consideration link speed, cost, or quality. RIP tables are dynamically generated by all participating routers and are updated regularly. All RIP-based routers will broadcast their entire RIP database to all neighboring routers every 30–90 seconds whether there has been any change in topology or not. The frequency of the broadcasts

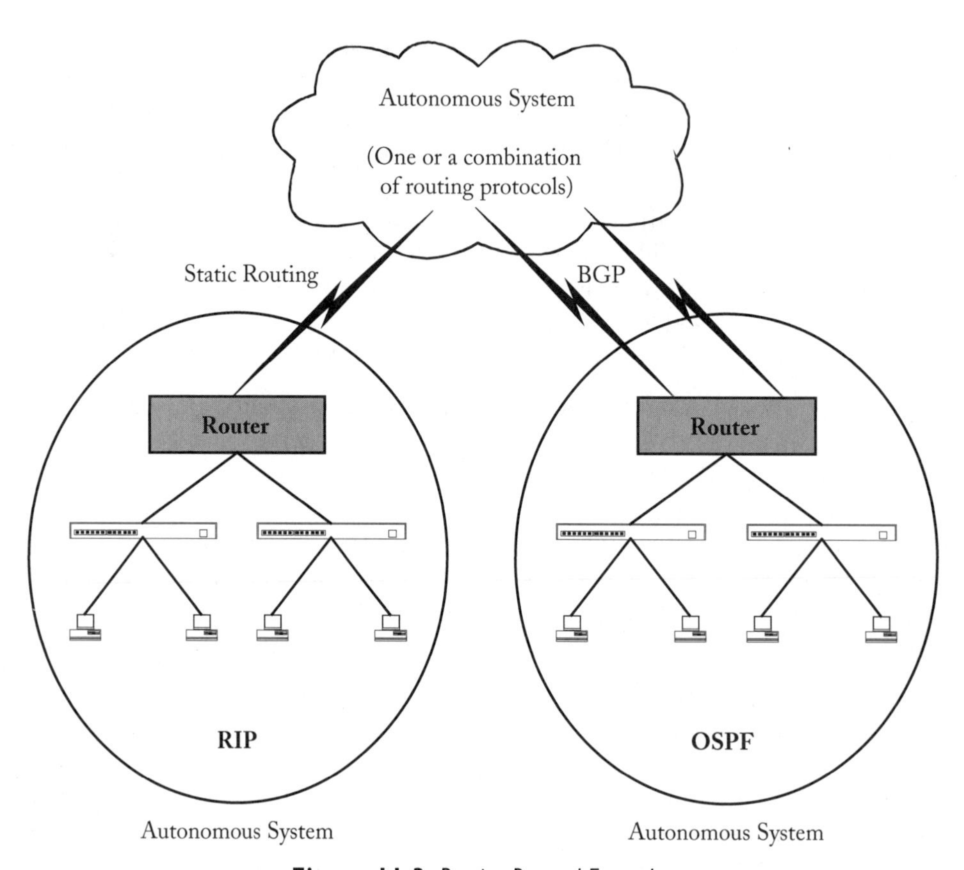

Figure 11.3 *Routing Protocol Examples*

and information contained therein depends on which Distance Vector routing protocol is used.

The IP-RIP broadcast interval is every 30 seconds and is conveyed between routers using UDP/IP packets. This is true for both IP-RIP version 1 and version 2. The Novell IPX-RIP broadcast interval is every 60 seconds and is conveyed between routers using IPX packets. (Also of potential interest is the Novell IPX-SAP broadcast, the interval for which is every 60 seconds and it is conveyed between NetWare servers using NCP/IPX packets.) The Cisco IP-IGRP broadcast interval is every 90 seconds and is conveyed between routers using UDP/IP packets. Table 11.2 compares these variations of the RIP routing protocol, their broadcast intervals, and the protocols used to convey the RIP messages across their respective networks.

> **Note:** Although the Novell Service Advertising Protocol (SAP) is not a routing protocol, it is often discussed in the same context. This is perhaps because NetWare servers are inherently routers, and they generate both SAP and RIP broadcast messages.

TABLE 11.2 A COMPARISON OF RIP-BASED ROUTING PROTOCOLS

RIP Variant	*Broadcast Interval*	*Protocol*
IP-RIP	30 seconds	UDP/IP
IPX-RIP	60 seconds	IPX
IPX-SAP	60 seconds	NCP/IPX
IP-IGRP	90 seconds	UDP/IP

When a topology change does occur, it may require from several seconds to a few minutes for the RIP messages to propagate across the internetwork before all of the router tables converge (i.e., when all router tables reflect the new information). As internetworks get larger, routing tables grow, and the time to converge takes longer. This also produces excessive traffic overhead, which consumes bandwidth that would otherwise be available for data. As routing tables grow they may exceed the maximum packet size supported by RIP, resulting in multiple packets for each RIP broadcast. In the event a single packet of a multi-packet RIP update is lost, the receiving routers discard all packets of that update. The receiving routers will have to wait for the next multi-packet RIP update. If enough RIP updates are not received, the advertised networks will automatically be purged from the routing tables—even though the networks may still be operating normally.

This sort of problem typically occurs only in large, heavily congested internetworks. Once the networks are purged from the tables the congestion usually subsides, which allows the RIP updates to propagate between routers again. Of course, once the routers receive the updates, the networks are added to their routing tables to indicate that those networks are again reachable, the congestion returns, and the cycle begins again. This phenomenon is sometimes referred to as *route flapping*. Several methods are available to suppress flapping, but the best alternative is to upgrade to a more robust routing update protocol such as OSPF.

In addition, since RIP is concerned only with minimizing router hops between networks, it does not always select the best path over which to forward packets. For example: A user on Host A wishes to access Host B using TCP/IP (with RIP). The fastest path would be from Router 1 over the three T1 links to Router 4. However, the path with the least number of intermediate hops is from Router 5 over the 56 kbps link to Router 6 (see Figure 11.4). RIP is a fine routing protocol for relatively small, simple hub-and-spoke connections, but RIP clearly is inadequate to meet the demands of large, diverse, growing internetworks.

Many large organizations have pushed RIP to its technical limits—in its extent, the number of users and subnets that can be effectively interconnected, and consumption of network bandwidth. In response to the need, Cisco Systems long ago developed

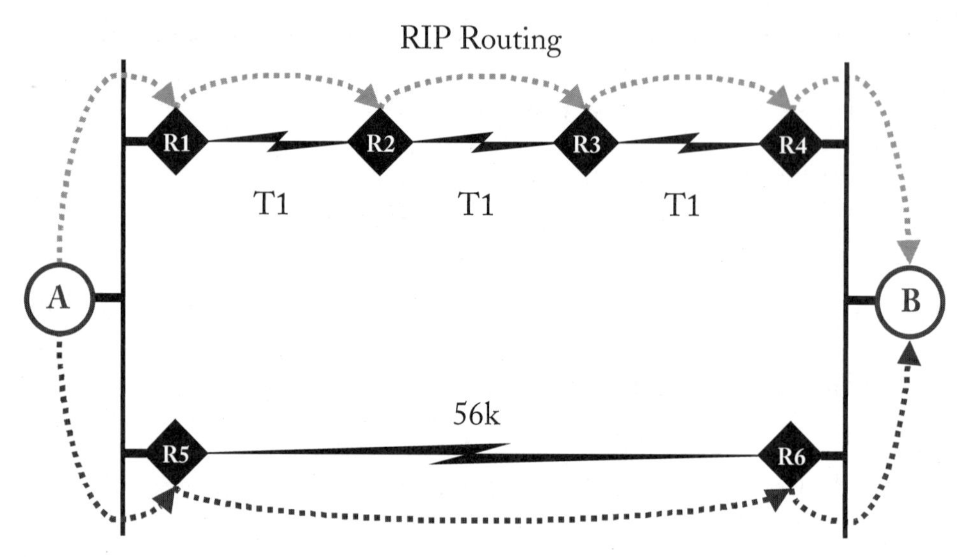

Figure 11.4 *Routing Information Protocol (RIP)*

IGRP as an alternative to RIP; the IETF developed OSPF; the ISO developed IS-IS; and Novell developed enhancements to their IPX protocol to address several of these issues. Today most router vendors support RIP, RIP-II, and OSPF for IP, and several router vendors also support Novell's IPX enhancements. Most Novell networks now support TCP/IP. If need be, it is possible to implement OSPF in the same internetwork with RIP, permitting a gradual transition from RIP over to OSPF.

OPEN SHORTEST PATH FIRST

OSPF is a Link State protocol that consists of a more complex algorithm based on link speed, cost, or quality. OSPF does *not* consider the number of router hops. All participating routers dynamically generate OSPF tables, but once the information is distributed to all routers in a given area, the only updates transmitted include changes to the network. OSPF-based routers do *not* broadcast their entire route database regularly; instead, OSPF dynamically monitors changes in the status of each of its network links and broadcasts any changes immediately. Rather than broadcasting the entire routing database, OSPF transmits only the changes, which consumes less bandwidth. OSPF routers do broadcast their presence every 30 minutes. This is a form of sanity-check, since OSPF routers do not make frequent contact with their neighbor routers (i.e., "Are we still in business?"). If there are no changes in link status, no additional broadcasts are made. This greatly reduces route management traffic overhead across the entire internetwork and optimizes internetwork link utilization.

To improve routing efficiency and optimize routing update requirements, OSPF supports the concept of *areas*. OSPF areas provide another level of hierarchical management, and are similar in application to the area codes used in the North American telephone system. The North American Numbering Plan (NANP) defines a hierarchical addressing scheme based on a ten-digit number. This ten-digit number is divided into the following hierarchy: area code (three digits), exchange code (three digits), and subscriber (four digits). With OSPF routing, an organization can divide its IP addressing scheme into areas, networks, subnets, and hosts. And whereas the NANP defines a ten-digit addressing scheme with a fixed number of decimal digits for each level, IP addressing defines a 32-bit addressing scheme that can support a range of bits allocated to each level. For example, the network may require 16 bits, the subnet field 8 bits, and another 8 bits for the host field. Or the network may require 16 network bits, 12 subnet bits, and 4 host bits. The ability to allocate varying numbers of network and host bits is controlled by the subnet mask. Because of this, IP addressing is far more flexible than the NANP. OSPF areas are defined by a seperate 32-bit value, adding even greater flexibility.

Further optimization can be achieved via the use of OSPF route summaries. This requires a bit of strategic planning and intelligent allocation of IP address space, but the benefit can be well worth the effort. A route summary is simply using just a portion of the leading bits in an IP number to identify a subset of a group of IP networks. This means allocating sequentially numbered blocks of IP address space. For example, networks 10, 11, 12, 13, 14 could be summarized as the "10s," and networks 20, 21, 22, 23, 24 could be summarized as the "20s." Each of these groups of networks would be allocated to a single OSPF Area so that routers can identify their locations more easily. This concept does get a bit more complicated when applied to 32-bit IP addressing, but that is beyond the scope of this text (see Figure 11.5).

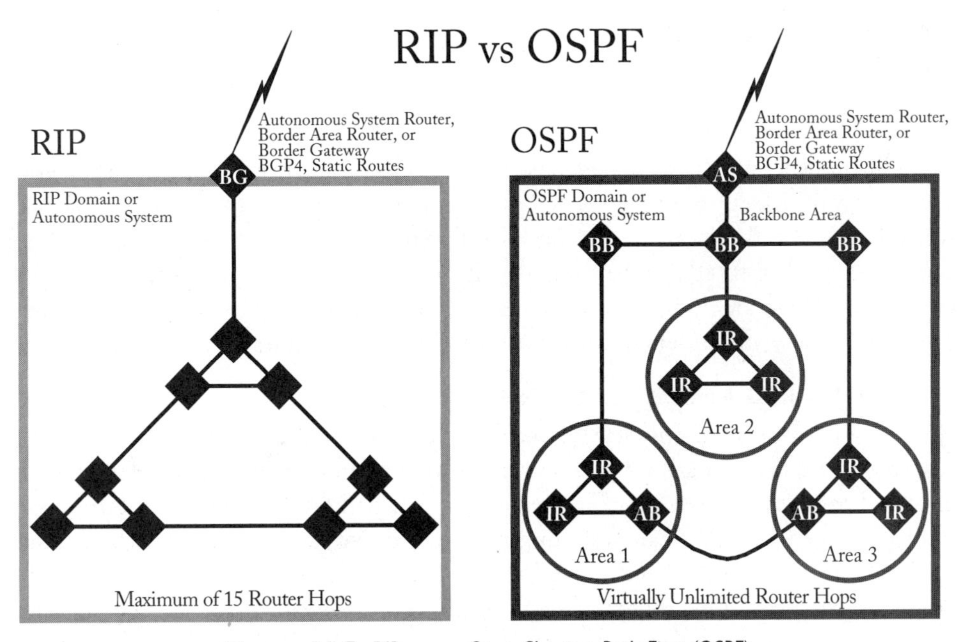

Figure 11.5 *RIP versus Open Shortest Path First (OSPF)*

Whereas the RIP architecture is very simple and limited to just fifteen router hops, the OSPF architecture can be complex and is virtually unlimited. Because of this increased complexity, where all RIP-based routers are equal, OSPF defines several different router functions based on the specific requirements of each task.

OSPF defines a hierarchy of routers or router functions. The same routers support these various OSPF functions; it's just a matter of configuring each function where it is required.

AS	Autonomous system router
BB	Backbone area routers
IR	Inter-area routers
AB	Area border router

Autonomous system routers provide routing between different autonomous systems and may use an Exterior Gateway Protocol such as BGP4, or static routes. Backbone area routers provide the interconnections between all areas within an autonomous system. Inter-area routers provide internetworking within a given area. And finally, area border routers provide direct area–area connections that circumvent the backbone area.

This is by no means a comprehensive explanation of OSPF and is intended only as a primer. Effective OSPF implementations require a great deal of study and forecasting to design an IP addressing scheme and area hierarchy that will efficiently support an organization's needs for many years. A solid understanding of IP addressing and subnetting is also essential. More information on OSPF may be found in rfe 1583.

ADDITIONAL FEATURES

In addition to forwarding packets between networks and subnets based on network addressing, routers also provide an array of features and options to better control traffic flow. Some of the more commonly used features include load sharing and/or load balancing, packet filters, and other security mechanisms such as network address translation. These are just a few of the many capabilities offered by today's routers and layer 3 switches—to understand them all can be a career in itself.

LOAD SHARING AND BALANCING

Routers are responsible only for forwarding packets on to the next leg of their journey toward their destination host. Some routing update protocols can calculate only one valid path at any given time. For example, if two paths are available to reach a given destination network, RIP can only support one or the other. However, OSPF can distribute the packet traffic across multiple paths simultaneously using a technique called load sharing. Some router vendors also support hardware-based load sharing that can be used regardless of routing update protocol.

There are a number of different load sharing methods, including per packet and per session. Per-packet load sharing simply alternates sending outbound packets over each available path, but without regard for packet size, upper layer protocols, or application performance requirements. Per-session load balancing requires the router to monitor each TCP or UDP connection, or session, and keep all packets associated with a given session on the same path for the duration of the session. This is done without regard for the duration of each session, the traffic load imposed by each session, or application performance requirements. No attempt is made to ensure that the traffic burden is balanced across the multiple network paths; this would be referred to as *load balancing* and requires a bit more sophistication than is afforded by simple load sharing.

PACKET FILTERS

Many routers offer additional features including the ability to restrict or prioritize packets based on their contents. Also referred to as an Access Control List (ACL), packet filters are a common method of securing a network from unwanted forms of traffic. Packets may be filtered based on a number of different criteria, including source or destination network protocol address, the type of upper layer protocol used, and even the application that generated the data contained in each packet. For example, a router could be configured to filter specific IP packets based on the source and/or destination addresses found in all IP packet headers.

The IP packet header also contains a protocol ID field that defines the upper layer protocol contained within the IP data field. Examples include the number "6," which represents TCP and the number "17," which represents UDP. Packet filters could be configured to block or forward packets containing either of these protocols. The TCP and UDP headers contain additional information in the form of a source port ID and destination port ID. Routers can identify this information and either block or forward specific packets depending on how the packet filter is configured. This is just a sampling of the packet-filtering capabilities provided by some routers.

NETWORK ADDRESS TRANSLATION

There are also situations where the need exists to translate packet addressing between two or more networks. This capability is commonly used to connect private corporate IP networks to the Internet. Because of security issues, IP address availability and other such concerns, most organizations implement *private* IP addressing within their private intranets. The IETF has set aside several address ranges specifically for use in private networks. Private IP addressing is specified in RFC 1918 (and updated in later RFCs). The addressing used on the Internet is referred to as *public* IP addressing. Public addresses are registered to a given organization, they may not be shared on multiple hosts, and they are typically not portable (i.e., if you switch ISPs, the public IP addresses remain with the ISP).

To fully understand Network Address Translation (NAT) requires a complete understanding of IP addressing and subnet mask calculations. As the name implies, NAT exchanges the IP addresses in IP packets as packets are forwarded from a private network onto the Internet, and vice versa. This allows an organization to implement whatever private addressing scheme they wish in their private intranet, and connect to the public Internet through a NAT gateway. The NAT gateway requires one or more public IP addresses that it will use to exchange for the private IP addresses used on the intranet. While private IP addresses are routable, they are *not* routed over the public Internet; they must first be translated into public addresses.

Technically, a basic NAT requires one public IP address for every private IP address it translates. Because obtaining this many public IP addresses is not often feasible, an extension to NAT has been developed called Port Address Translation (PAT). PAT allows NAT to translate many private IP addresses using just one public IP address or a small pool of public IP addresses. Most NAT implementations include PAT, and some vendors (specifically Cisco) refer to it as Network Address and Port Translation (NAPT).

The ports referred to are again found in TCP and UDP headers. In this context, the term "port" does not refer to a physical cabling interface; instead, a TCP/UDP port is a software abstraction referring to numbers that represent different network programs, applications, or services. Each application in a TCP/IP network uses a specific port number. TCP/IP applications include the World Wide Web (http—port 80), and the File Transfer Protocol (ftp—ports 20 and 21). The source and destination port fields in TCP/UDP each consists of 16 bits, which means there are 65,536 unique port numbers. The Port Numbers from 0 to 1023 are called the Well-Known Ports and support most of the programs, applications, and services we all use on a daily basis.

When a host generates a packet to request a service on another host, the destination port number must identify the service desired (e.g., http 80). For each new service requested, the host must use an available source port number in the range from 1024 to 65,535. For example, the first http request will have a destination port number of 80 and a source port number of 1024. The second request will have a destination port number of 80 and a source port number of 1025. The third request will have a destination port number of 80 and a source port number of 1026, and so on. Port Address Translation also utilizes unused port numbers in this range to perform its functions.

NETWORK OPERATING SYSTEMS AND PROTOCOLS

Most network operating system (NOS) vendors have developed their own network protocols, typically deriving them from IP or XNS. These vendors have also developed their own routing protocols, usually based on RIP, yet modified enough to prevent interoperating with other implementations of RIP used by other vendor's protocols.

Most multi-protocol routers can support Novell's IPX protocol. Such routers can also forward IPX packets between a Novell NetWare server on an Ethernet LAN and a NetWare client PC on token ring LAN (or any other standard LAN). But a router cannot forward packets between either IPX host and another host supporting only the TCP/IP protocol (such as a Unix host). Using only routers, network communications can only be provided directly between devices that use the same protocols.

Most computer operating systems now provide support for protocols other than their native protocols. Novell's original protocol suite consisted of Internetwork Packet Exchange/Sequenced Packet Exchange (IPX/SPX), Novell's own implementation of the Routing Information Protocol (RIP), and the NetWare-specific Service Advertising Protocol (SAP) and NetWare Core Protocols (NCPs). Novell now supports NetWare file and print sharing, as well as Internet applications over both IP and IPX. A special IP-IPX gateway is required to provide access to IPX-based Internet applications from the TCP/IP-based Internet. Of the NetWare installations that still exist, most have migrated to the later versions of NetWare that provide native TCP/IP support. Novell also provides direct support for Unix hosts, including Sun Microsystems' Network File System (NFS).

Microsoft operating systems, including everything from Windows for Workgroups (3.11) to Windows 95/98/Me, Windows NT, Windows 2000, and XP, all support Windows file and printer sharing over IP and IPX, as well as the non-routable NetBEUI protocol. Microsoft supports Internet applications over IP only. Both Microsoft and Novell also support AppleTalk clients. The many flavors of Unix provide support for IP and most of the Internet utilities, and some also include support for special utilities and protocols such as NFS. Unix support for Microsoft and Apple networking is available through third parties.

Originally, network operating system vendors offered no choice of network protocol or routing update protocol. If Novell NetWare was selected, you got IPX/SPX, RIP, and SAP. If a Unix system was selected, you got the suite of TCP/IP protocols and RIP. For many years TCP/IP users have had an array of choices: RIP-I, RIP-II, OSPF, Cisco's IGRP, and now EIGRP, or the ISO's IS-IS. Users of IPX and AppleTalk protocols also have alternatives; Novell has combined IS-IS-based routing and the SAP service advertising function into the NetWare Link Services Protocol (NLSP), and Apple developed the AppleTalk Update-based Routing Protocol (AURP). I apologize if the alphabet soup is getting a bit thick, but this is the language of protocols.

SERVER-BASED INTERNETWORKING

All TCP/IP protocol stacks inherently perform basic routing functions, but several host and file server operating systems include advanced routing features and even multiprotocol support. All versions of Unix, Microsoft Windows (95/98/Me and NT/2000/2003/XP), as well as the later releases of Novell NetWare provide several internetworking options. While these solutions may appear to provide the functionality of more expensive products, they are not without limitations. Server-based internetworking is an inexpensive solution if you already own the server, but performance, reliability, flexibility, and expansion are concerns (see Figure 11.6).

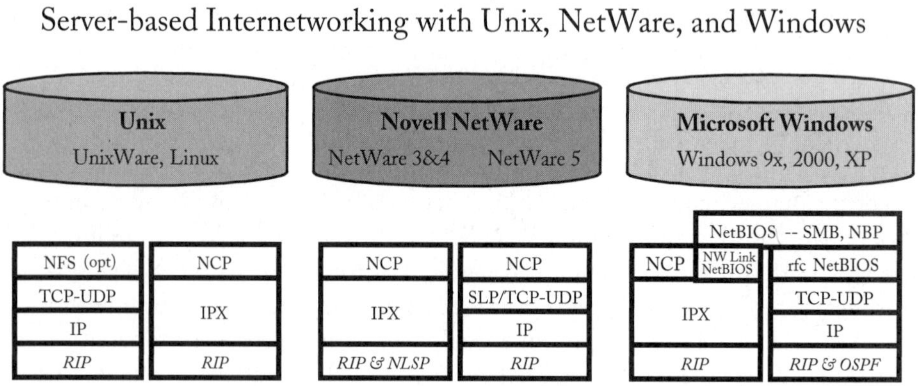

Server-based Internetworking with Unix, NetWare, and Windows

Unix
UnixWare, Linux

Novell NetWare
NetWare 3&4 NetWare 5

Microsoft Windows
Windows 9x, 2000, XP

NFS (opt)
TCP-UDP
IP
RIP

NCP
IPX
RIP

NCP
IPX
RIP & NLSP

NCP
SLP/TCP-UDP
IP
RIP

NetBIOS -- SMB, NBP
NCP
NW Link NetBIOS
IPX
RIP

rfc NetBIOS
TCP-UDP
IP
RIP & OSPF

Figure 11.6 *Server-based Internetworking*

Support for TCP/IP has been available in most Unix systems since the mid-1980s. IP routing is supported via the route daemon: *routed* (pronounced, "route-dee"). To provide simpler file and printer sharing capabilities between Unix hosts, Sun Microsystems developed the Network File System. And after purchasing AT&T's Unix Systems Labs, Novell integrated support for IPX and NetWare functionality into their newly acquired Unix operating system and called it UnixWare. Novell eventually sold the operating system to Santa Cruz Operations (SCO), now Caldera International, who still markets and supports the product under the name UnixWare.

For many years Novell supported only its own IPX protocol with RIP routing. Following the release of NetWare v.3.11, Novell offered support for IP along with its native IPX protocols. The additional protocol support was provided via optional NetWare Loadable Modules (NLMs). Network NLMs are server-based programs that include support for TCP/IP and several utilities, including IP Tunneling (encapsulation of IPX inside TCP/IP), an FTP server, Sun NFS, AppleTalk, ISO OSI (as if anyone cares), IBM SNA/SAA, and source route bridging for token ring installations that support IBM host systems.

NetWare/IP was released after NetWare v.4 and provided some improvements over the original IP Tunneling, but it still encapsulated IPX and had performance problems. The NLSP provides IS-IS-based routing for IPX and integrates the SAP functions. These advances greatly improved the performance and efficiency of IPX internetworking, but were obviously "too little too late" for IPX. With the release of NetWare v.5, Novell shifted its focus to provide native support for TCP/IP. This is accomplished through the use of the Service Locator Protocol (SLP), published as IETF RFC 2165 (among others). (See Figure 11.6.)

Since the days of Microsoft's ancient MS Net, LAN Manager, and Windows for Workgroups, Microsoft networking has relied on the NetBIOS protocol. NetBIOS has many drawbacks: it is a broadcast-based non-routable protocol, it requires a unique fifteen-character name for every computer in the network, and it does not scale well to support large enterprise internetworks. To make NetBIOS routable, it must be encapsulated inside a routable protocol such as TCP/IP. Microsoft achieves this by using NetBIOS over TCP/IP (NBT).

With the release of Windows NT v.3.x, support was added for Microsoft Networking using both NetBIOS over TCP/IP (with static routing only), and NetBIOS over IPX (with RIP routing). The Windows Internet Name Service (WINS) was developed to provide IP-Address-to-NetBIOS-name resolution for routing Microsoft Networking sessions over IP internetworks. Think of WINS as a Domain Name Service (DNS) for NetBIOS names (rather than Internet host names). WINS has always been somewhat unreliable and its database tends to become corrupted easily.

With Microsoft's release of Router Remote Access Service (RRAS) support for OSPF routing was added to NT Server v.4. Microsoft also released NT Domains, which require an array of NT servers to support the NT domain infrastructure—a Primary Domain Controller (PDC) and one or more Backup Domain Controllers (BDCs). With the release of Windows 2000 (briefly known as NT5), Microsoft has replaced the original NT Domains system with its new Active Directory (AD) service. Active Directory relies on domain controllers, but instead of defining a primary/secondary relationship, all AD domain controllers are peers. Windows 2000 also eliminates the requirement for NetBIOS (although some NetBIOS artifacts are still there). And if backward compatibility is required to support earlier versions of Windows, then NetBIOS over TCP/IP must be maintained. In either case, Microsoft Networking using TCP/IP is routable. The new method is more efficient than traditional NBT.

CHAPTER SUMMARY

Routers come in several forms, including purpose-built devices or software installed on a server or host computer system. The purpose-built device may be a large multi-slot chassis running a sophisticated operating system that supports numerous features and every LAN and WAN interface available, or a small plastic device with tow to eight ports, and limited features designed for the small office/home office. The server software must be written to support your particular computer platform and operating system, and is limited by the expandability and performance of the computer. There are many network protocols available, and most are routable; however, not all routers support all protocols. Routable network protocols support one or more routing update protocols. Being the most popular protocol on the planet, TCP/IP offers the most choices in this area. The subject of routers and protocols is quite complex, requiring a great deal of study on a wide array of topics. The purpose of this chapter has been to provide the reader with a basic foundation of internetworking technologies.

REVIEW QUESTIONS

1. Routers operate at what layer of the OSI model?
2. What are the differences between routers and bridges?
3. What the difference between dynamic and static routing?
4. Identify four interior gateway protocols.
5. Define the difference between routable and non-routable protocols.
6. Identify four ways in which OSPF differs from RIP.
7. Identify four advantages to using routers.
8. Define the following acronyms:

IAB	IETF
RFC	OSPF
RIP	IGRP
IGP	IS-IS
BGP	NLSP

9. What is an OSPF route summary?
10. Identify four problems with RIP-based routing.

Cable and Cabling Standards

OBJECTIVES

After completing this chapter, you should be able to:

- Describe each of the primary types of media and identify applications for each.
- Explain the differences between STP, STP-A, ScTP, and FTP.
- Identify the key differences between UTP and STP.
- Explain the gauges and size measurements of copper conductors and optical fiber.
- Describe the color-coding and pin-out specifications for most UTP cables.
- Explain the difference between the pin assignments of T568-A and T568-B.
- Define the key electrical characteristics for Category 3, 5, and 5E UTP.
- Describe multi-mode fiber and single-mode fiber.
- Define ANSI/TIA/EIA 568-A.

INTRODUCTION

This chapter discusses the wide variety of network media available, both copper and glass, and the physical and electrical or optical properties of each. You will learn about the advantages and disadvantages of each type of media, as well as the standards that have been developed to define them. In addition to the various media specifications, there are several structured cabling systems standards required for any modern cable plant installation.

CABLING OVERVIEW

Cabling technology is a completely separate facet of networking. An important distinction in the terminology of cabling is the term *cabling* itself. Cabling refers to a complete system of media, connectors, and components, whereas the term *cable* refers just to the media. In some cases special cabling systems have been engineered just to satisfy the unique requirements of a specific network technology, and in other cases a network technology has been developed to take advantage of an existing cabling system. For example, 10Base5 Ethernet specifies a 50-Ohm thick coaxial cable that has no practical application other than 10Base5. 10Base-T was engineered to support the existing UTP cabling used in telephony (i.e., DIW-24), and 100Base-TX and 1000Base-T were both engineered to support the existing ANSI/TIA/EIA cabling standards for Category 5 UTP.

There are many diverse forms of network cabling available. Over the years many have enjoyed some degree of popularity only to later fade into obscurity. Each has certain advantages, in the ability either to provide greater bandwidth, to carry signals over greater distances, or to provide simpler and more cost-effective installations, moves, adds, and changes. For example, while coaxial cabling does provide greater bandwidth over greater distances than UTP, UTP is more popular for data networking due to its support of structured cabling and low cost.

Conductor size is based on American Wire Gauge (AWG), which is the U.S. standard for measuring the size of non-ferrous conductors. Gauge measures the diameter of a conductor using a retrogressive scale. In other words, the higher the AWG number, the smaller the diameter of the conductor. This strange practice emerged as a result of early wire-manufacturing methods, where each conductor was repeatedly drawn through a wire machine to achieve the desired diameter. For example, a 22-gauge wire was drawn through the machine 22 times; therefore, it is not as thin as a 24-gauge wire, which was drawn through the machine 24 times.

Most UTP used today meets the standards established by the ANSI/TIA/EIA (often referred to as just the TIA/EIA). These cabling standards define several categories of UTP, STP, and optical fiber, in addition to *structured cable plant* design, installation, and administration practices. A structured cable plant is a wiring infrastructure installed and certified to comply with a set of standards defining expected electrical properties. Structured cable plants are usually configured in a star-wired topology to facilitate administrative and technical issues such as moves, adds, and changes. UTP, STP, and fiber are only media used to construct structured cable plants.

COAXIAL

Coaxial cabling was used in some of the earliest LANs. Many of these early LANs were based on proprietary technology, and they often required proprietary media. Coaxial cable, often called simply *coax*, comes in many different varieties and sizes, each having distinctly different electrical characteristics. Many coaxial cables support a very large bandwidth and can therefore provide very high data-carrying capacity.

Coaxial cable consists of a solid or stranded copper center conductor conforming to some specific diameter. An insulator or dielectric, again conforming to some specific diameter, surrounds the center conductor. Coaxial insulators may be solid plastic material, closed cell foam, or a hollow tube. One or more conductive shields then surround the insulator. Shielding material usually consists of a metal braid, a foil wrap, or both. This entire assembly is then wrapped by another insulating jacket, all of which forms a series of concentric circles surrounding a single center conductor. Most coaxial cabling is identified by a Radio Guide (RG) number, such as RG-58. The numbers themselves are simply a standards reference and contain no inherent significance (see Figure 12.1). Table 12.1 shows examples of coaxial cables and their characteristic impedance, physical construction, and most common applications.

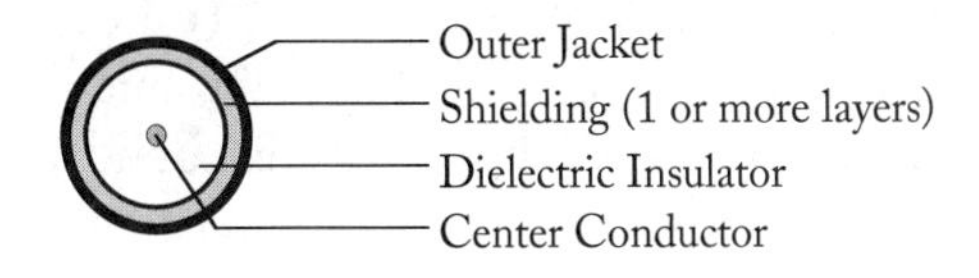

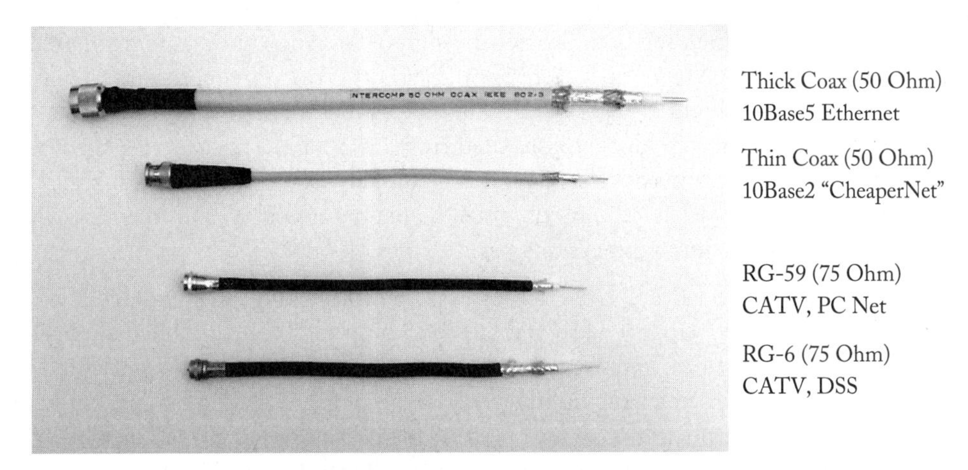

Figure 12.1 *Coaxial Cables*

TABLE 12.1 COAXIAL CABLE TYPES, PROPERTIES, AND APPLICATIONS

Media	*Impedance*	*Details*	*Applications*
RG-58	50–53 Ohms	Solid copper center conductor	Mistakenly used for 10Base2
RG-58 a/u	50 Ohms	Stranded, tinned copper center	10Base2 Ethernet
RG-58 c/u	50 Ohms	Stranded, tinned copper center	10Base2 Ethernet
RG-59	75 Ohms	Solid copper center conductor	CATV, PC Net
RG-6	75 Ohms	Solid copper center, quad shielded	CATV, Satellite
RG-62	93 Ohms	Solid copper center conductor	IBM 3270, Arcnet

TWINAX

Twinax is like coaxial cable but with a pair of parallel center conductors inside a common insulator, surrounded by a metallic shield, wrapped inside an insulating jacket. Twinax has been used in applications such as cable television, and in IBM's iSeries and AS/400, primarily to provide backward compatibility with the company's older System 36/38 line of mini-computers. More recently, twinax has been considered for use in short-haul Gigabit Ethernet applications (see Figure 12.2).

SHIELDED TWISTED PAIR

There are many varieties of shielded cabling, including shielded twisted pair (STP), screened twisted pair (ScTP), and foil twisted pair (FTP). All varieties consist of two or more twisted copper pairs enclosed in a metallic shield, wrapped in an insulating jacket. The shielding may be either a heavy metallic braid, as is typical for STP; a thin metallic mesh in the case of ScTP; or simply a foil wrap in the case of FTP. In all cases, proper electrical grounding is essential (see Figure 12.3).

The purpose of the shielding is to protect the signal-carrying conductors from external sources of Electro-Magnetic Interference (EMI) and radio frequency interference (RFI); however, the shield also acts as an antenna. Ambient EMI/RFI noise induces a current flow into the shield, therefore requiring a proper electrical (earth) ground. If the shielding is improperly grounded, it will actually exacerbate the problem it is intended to prevent! The shield integrity must be maintained from end to end, as well as through all cabling interconnections.

The most common STP are IBM Types 1, 2, 6, and 9. The original IBM STP cabling was certified to support a spectral bandwidth up to 20 MHz. Shielded twisted pair-A (STP-A) is a modification of the original specifications. STP-A is certified to support an increased spectral bandwidth up to 300 MHz using the same cable, but with improved connectors and components. The new connectors are interoperable with the old, but mixing STP and STP-A components reduces the cable's capabilities to

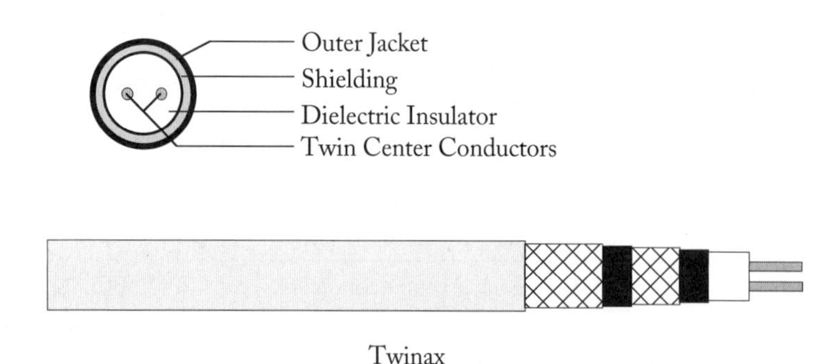

Figure 12.2 *Twinax Cable*

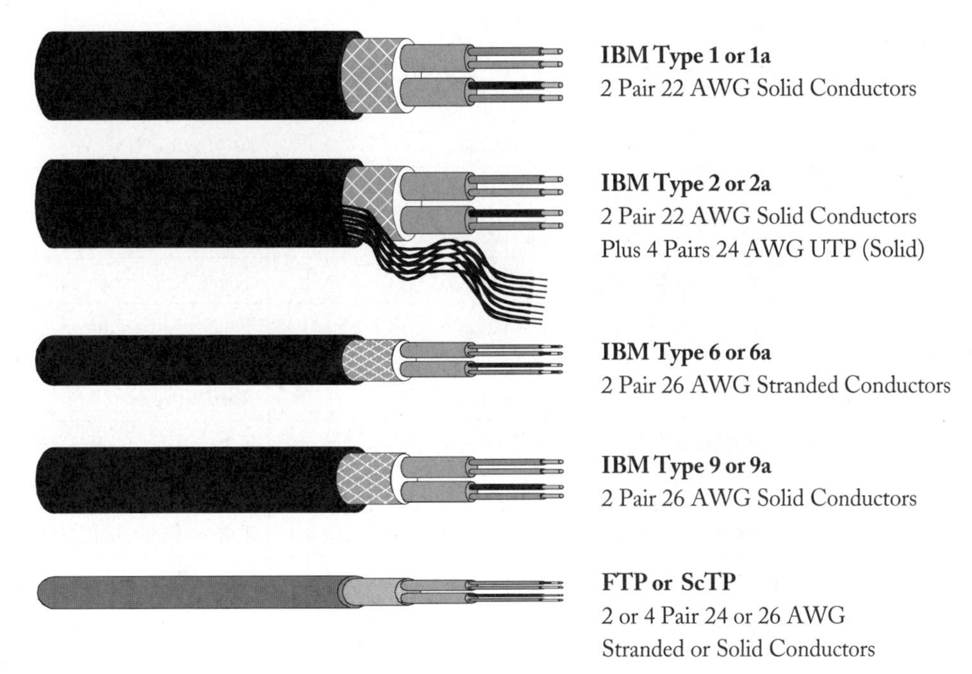

Figure 12.3 *Shielded Twisted Pair*

those of the original STP specifications. The purpose of STP-A is to support 100 Mbps High-Speed Token Ring, and Gigabit Token Ring.

The above terminology may cause problems in the United Kingdom and elsewhere. In the United Kingdom, *shielding* is referred to as *screening*, so that all STP is called screened twisted pair. Because of this it is difficult to consistently differentiate UK STP from ScTP. Also in the United Kingdom, *electrical grounding,* or *earth ground,* is referred to as *earthing* or simply *earth.*

Table 12.2 shows examples of shielded twisted pair cables and their characteristic impedance, physical construction, and most common applications.

UNSHIELDED TWISTED PAIR

UTP is by far the most common cabling used in data networking today. There are several varieties of UTP currently in use, and several others that have been discontinued. For years UTP has been the primary media used for telephone systems in homes and businesses. With the emergence of 10Base-T Ethernet, UTP became the primary media for local area networking everywhere. 10Base-T specifies support for two pairs of 22-26 AWG UTP media with a differential characteristic impedance of 100 Ohms ±15 Ohms. These specifications are met by ordinary UTP telephone wire, commonly referred to as Type-D Inside Wire (DIW), which is available in 22,

TABLE 12.2 SHIELDED TWISTED PAIR CABLE TYPES, PROPERTIES AND APPLICATIONS

STP	*Impedance*	*Details*	*Applications*
Type 1	150 Ohms @ 3–20 MHz ±10%	2-pair, 22 AWG solid	Token Ring, CDDI
Type 2*	150 Ohms @ 3–20 MHz ±10%	2-pair, 22 AWG solid	Token Ring, CDDI
Type 6	150 Ohms @ 3–20 MHz ±10%	2-pair, 26 AWG stranded	Token Ring patch cord
Type 9	150 Ohms @ 3–20 MHz ±10%	2-pair, 26 AWG solid	Token Ring, reduced distance
STP-A	***Impedance***	***Details***	***Applications***
Type 1a	150 Ohms @ 300 MHz	2-pair, 22 AWG solid	Token Ring, CDDI
Type 2a*	150 Ohms @ 300 MHz	2-pair, 22 AWG solid	Token Ring, CDDI
Type 6a	150 Ohms @ 300 MHz	2-pair, 26 AWG stranded	Token Ring patch cord
Type 9a	150 Ohms @ 300 MHz	2-pair, 26 AWG solid	Token Ring, reduced distance

*Types 2 and 2a include an additional 4-pairs of Type 3 UTP (outside the shield, under the jacket).

24, and 26 AWG. New cable plant installations generally rely on TIA/EIA standard Category 3, Category 5, or Category 5E UTP.

There are many references to 10Base-T that claim the standard requires Category 3 UTP, but this is not technically accurate, nor chronologically possible. The IEEE 802.3i standard that defines 10Base-T Ethernet was published several years before the TIA/EIA standard for Category 3. While 10Base-T will certainly operate over Cat3, Cat4, Cat5, or Cat5E, the 10Base-T specifications merely require ordinary phone wire, such as DIW-24. Most DIW-24 is comparable to Cat3, but may fall short of several Cat3 electrical specifications.

An unshielded twisted pair consists of two insulated copper conductors twisted around each other. The insulation covering each conductor is color-coded, typically according to the TIA/EIA specifications: Pair 1 is blue/white-blue, Pair 2 is orange/white-orange, Pair 3 is green/white-green, and Pair 4 is brown/white-brown (or *slate*) (see Figure 12.4).

To confine the electromagnetic field generated by each pair, the conductors of each pair are twisted around each other at specific intervals. Although some UTP cables may support from 2 to 25 pairs, today most UTP intended for high-speed LAN applications consists of 4 pairs. The electromagnetic field generated by each pair can cause the induction of signals from one pair to another. This is called *crosstalk* (XTALK), and to minimize crosstalk each pair is twisted at a different interval. The copper conduc-

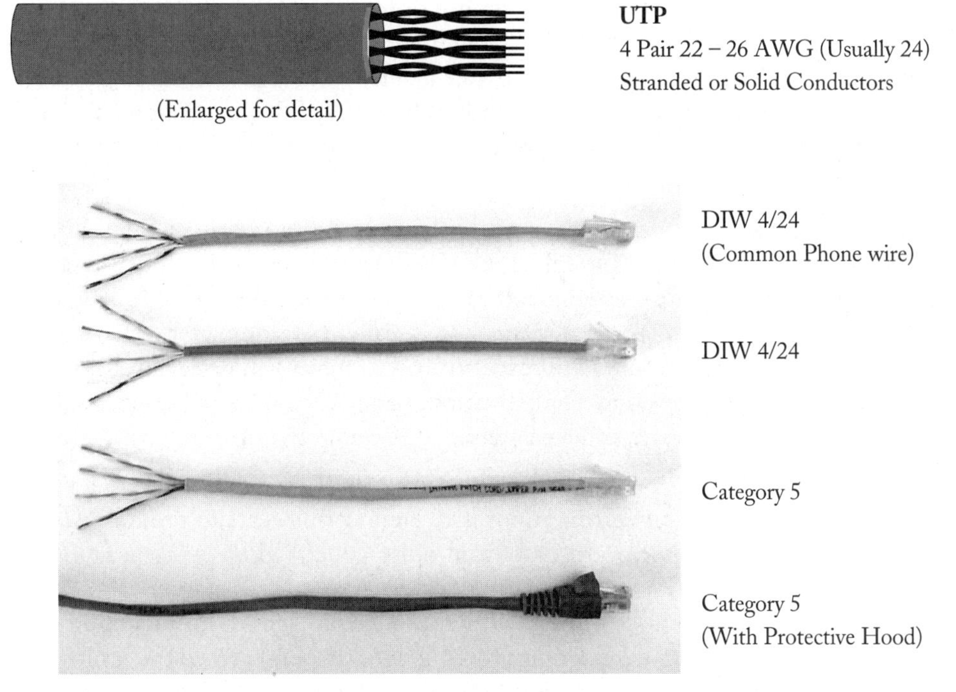

Figure 12.4 *Unshielded Twisted Pair*

tors are electrically balanced to equalize the current, or load, between elements of the circuit. For UTP cabling, balance is everything. Superior balance optimizes the cable, thereby providing stronger signals over greater distances while obviating the need for shielding and its related expenses and problems.

A mundane but crucial issue involving any UTP cable plant installation is the pair assignments (or pin-outs) of the connectors. Over the years several different UTP connector pin-out arrangements have been used. One of the earliest is called the Universal Service Order Code (USOC), which is used for most telephony cable plant installations. Because of the way pairs are split at each connector, the USOC scheme is among the poorest possible wiring configurations for data transmission and should therefore be avoided. The TIA/EIA has defined two similar pair assignment specifications for use with all categories of UTP cabling: TIA 568-A and TIA 568-B (see Figure 12.5 on the inside back cover).

Not to be confused with the ANSI/TIA/EIA 568-A cabling standard (or its European counterpart, ISO/IEC 11801), the standards specifying cable connector pair assignments are defined within the standard and unfortunately share very similar nomenclature. TIA 568-A or TIA 568-B (aka, T568-A and T568-B, respectively) define the connector pair assignments for 8-position modular jacks and plugs. Electrically, the pair assignments of the 8-position modular connector (commonly referred to as RJ-45) are the same. The only difference between T568-A and T568-B is that the position of pair 2 and pair 3 are swapped (the orange and green pairs). Although it is currently the least popular, T568-A is compatible with both USOC and ISDN pin assignments, and it is the only configuration recognized by the U.S. government (see Figure 12.6 on the inside back cover).

It is expected that T568-A will eventually become more prevalent than T568-B. T568-B is by far the most commonly used today, probably because T568-B is equivalent to the AT&T 258-A pin assignments traditionally used in AT&T cable plant installations. Both standards are electrically equal, so it really does not matter which standard you choose—just pick one and use it exclusively. Whatever you decide, just do *not* try to mix and match T568-A and T568-B components! Doing so will result in crossed pairs, a defective cable plant, and a tremendous amount of wasted time and effort.

UTP is not without its problems and limitations. It is not an ideal medium for high-frequency transmissions, requiring innovative engineering feats to produce network adapters that can reliably discern data from noise generated by its own data transmissions. ANSI/TIA/EIA standards for all categories of UTP restrict cable runs to a maximum of 100 meters (328 feet). While an adequate length for most horizontal cable runs, vertical runs in tall buildings (over about 20 floors) exceed the limits of the standard. UTP also makes a poor choice for cabling between buildings. In addition to distance limitations, copper media is sensitive to harsh environments, can create ground loops, and tends to attract lightning. The big advantages of UTP are that it is cheap and easy and supports structured cable plants.

OPTICAL FIBER

Fiber overcomes most of the problems inherent in all forms of copper cabling and provides many advantages. Fiber comes in several varieties, including multi-mode and single-mode, and each is available in several sizes. Most LAN standards support multi-mode fiber over distances up to 2000 meters (2 km, or 6,560 feet). Several manufacturers have developed hardware to support single-mode fiber; however, the only LAN standard currently supporting single-mode fiber is 1000Base-LX Gigabit Ethernet and is compliant with IEC 60793-2:1992 (see Figure 12.7).

Whereas UTP conductor size is expressed in AWG, optical fiber sizes are expressed in micrometers (μm), such as 62.5/125 μm. The first number (62.5 μm) represents the diameter of the fiber core, while the second number (125 μm) represents the diameter of the cladding material that surrounds the core. Light wavelengths are expressed in nanometers (nm), such as 1300 nm. The visible light spectrum consists of a range of wavelengths from approximately 400 nm to approximately 700 nm (or 4×10^{-7} m to 7×10^{-7} m).

Most optical fiber network technologies require 2 strands of glass, one used to transmit and the other to receive. Networks using multi-mode fiber generally require wavelengths from 770 nm to 1300 nm, and support distances up to 2 km. Single-mode fiber applications such as Gigabit Ethernet specify wavelengths in the range of 1270 nm to 1355 nm over distances up to 5 km. Achieving the maximum distance usually requires network components to be configured for full duplex operation. Telephony and carrier applications can achieve much greater distances of 30 km to 40 km, depending on the equipment.

Since fiber carries light rather than electricity, it generates no electromagnetic fields that can cause the induction of signals from one pair to another, or crosstalk. Fiber also eliminates the possibility of a multitude of other noise-related problems common in copper networks. In addition, because optical fiber is glass and glass is non-conductive, it is not sensitive to harsh environments, cannot create ground loops, nor will it attract lightning. The big advantages of fiber are distance, bandwidth, immunity to noise, and increased security. Fiber also supports structured cable plants.

CABLING STANDARDS

As with LAN technologies, cable and cabling technologies exist that are proprietary, de facto standards, and industry standards. As the cabling industry began to develop its own standards, the LAN industry was quick to embrace them, eventually displacing virtually all proprietary cable plants and most de facto cable plants. One of the earliest de facto standard cable plants is the AT&T Systimax Premise Distribution System, known as PDS 110, which defines detailed specifications for a cable, connectors (plugs and jacks), installation requirements, and testing and certification procedures for the entire plant. Other proprietary PDS examples include the IBM Cabling System, Digital Open DECconnect, and Nortel Integrated Building Distributed Network (IBDN). The earlier purely proprietary systems are not worth mentioning.

With so many possibilities, most of which were not interoperable, organizations looking to develop a long-term cable plant strategy were forced to make some hard choices. Some cabling choices excluded the possibility of ever switching to an alternative network technology. What made things even worse was the range of quality of the product being sold by the numerous cabling manufacturers.

As a means to distinguish among the hundreds of different four-pair copper cables available on the market, in the late 1980s a company by the name of Anixter developed their own cable Levels Program. In 1992, the cabling industry adopted the Anixter Levels Program specifications as a new standard. The Levels Program began as a component specification for UTP cable and connecting hardware. It later evolved into an end-to-end channel specification with three distinct levels of channel performance, Anixter Levels Channel (ALC) 5, 6, and 7. The Levels Program continues to evolve, and now includes active (Mbps-based) testing, called Levels XP. Anixter is singled out because of its contributions to the cabling standards on which the entire industry now relies. The ANSI/TIA/EIA 568-A cable categories are based on the work pioneered by Anixter.

The National Electrical Code (NEC), is the wiring standard most widely supported by local building licensing and inspection officials in the United States. Among other details, the NEC requires the use of *plenum*-rated cable in air-return spaces. In most commercial construction, the space above suspended ceilings is used as one huge duct to provide a return air path for the heating, ventilation, and cooling (HVAC) system. Network cables are frequently installed in this space. Ordinary cable is typically constructed of Polyvinyl Chloride (PVC), which when burned (or heated to a sufficiently high temperature) will release toxic fumes that can kill. Obviously, installing PVC cable in a plenum return air space is a problem, and plenum-rated cable is the solution. Plenum cable is often constructed of Teflon™, Polyvinylidene Diflouride (PVDF), or some other material that does not release toxic fumes when burned. Plenum cable is also a bit more expensive than PVC cable, and because it tends to be less flexible than PVC it is more difficult to work with.

ANSI/TIA/EIA STANDARDS

The American National Standards Institute, Telecommunications Industry Association, Electronics Industry Alliance (ANSI/TIA/EIA) is responsible for defining the cabling standards used in virtually all network cable plant installations today. In addition to standards defining the electrical properties of the raw media, this standards body has also developed performance specifications for the entire cable channel, structured cabling systems, and more. The collection of cabling standards includes:

- 568-A Structured Cabling System
- 569-A Design Considerations
- 606 Administration Standard for the Telecommunications Infrastructure of Commercial Buildings
- 607 Commercial Building Grounding and Bonding Requirements for Telecommunications

ANSI/TIA/EIA 568-A Structured Cabling System

The ANSI/TIA/EIA 568-A standard defines six subsystems of a structured cabling system, including the Building Entrance, Equipment Room, Backbone Cabling, Telecommunications Closet, Horizontal Cabling, and Work Area (see Figure 12.7).

1. Building Entrance

Building entrance facilities are the points at which an external network carrier's cabling interfaces with the intra-building backbone cabling. The physical requirements of the network interface are defined in the TIA/EIA 569-A standard.

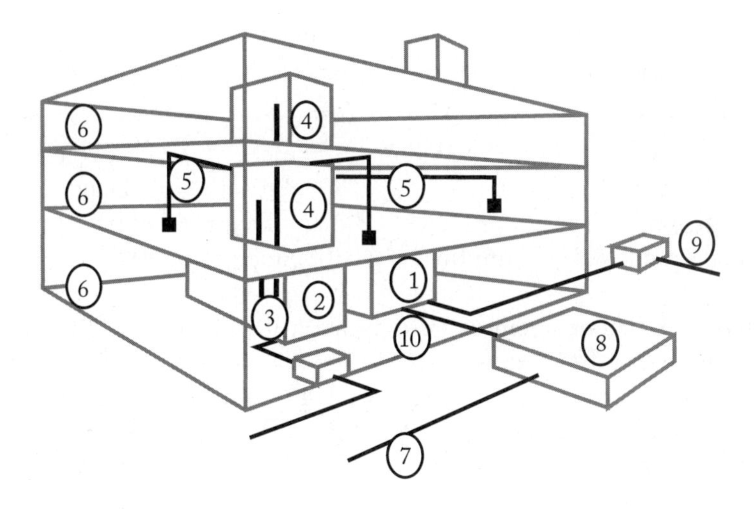

1. Telco Entrance Facility	6. Work Area
2. Telcom Equipment Room	7. Telco Conduit
3. Vertical Backbone	8. Telco Manhole
4. Telecom Closet	9. Interbuilding Backbone
5. Horizontal Cabling	10. Entrance Conduit

Figure 12.7 *TIA/EIA 568-A Structured Cabling System*

2. Equipment Room

Equipment rooms usually house more complex network hardware than telecommunications closets. An equipment room may provide any or all of the functions of a telecommunications closet. Equipment room design specifics are defined in the TIA/EIA 569-A standard.

3. Backbone Cabling

Backbone cabling provides the interconnection between telecommunication closets, equipment rooms, and entrance facilities. It consists of the backbone cables, intermediate and main cross-connects, mechanical terminations, and patch cords or jumpers used specifically for backbone-to-backbone cross-connection. This includes:

- Vertical riser connections between floors
- Cables between an equipment room and building cable entrance facilities
- Inter-building cables on a campus

The following list identifies the types of cabling recognized for use in backbone cable plants and the maximum backbone distances supported by each.

Cabling Types Recognized	**Maximum Backbone Distances**
100 Ohm UTP (22 or 24 AWG)	800 meters (2625 ft) Voice*
150 Ohm STP	90 meters (295 ft) Data*
Multi-mode 62.5/125 µm fiber	2000 meters (6560 ft)
Single-mode 8.3/125 µm fiber	3000 meters (9840 ft)

*Note: Backbone distances are application-dependent. All LAN technologies are limited to 100 meters of UTP! The maximum distances specified above are based on voice transmission for UTP and data transmission for STP and fiber. The 90-meter distance for STP applies to applications with a spectral bandwidth of 20 MHz to 300 MHz. A 90-meter backbone distance also applies to UTP at spectral bandwidths of 5–16 MHz for Cat 3, 10–20 MHz for Cat 4, and 20–100 MHz for Cat 5. The addition of 5 meters of patch/station cord at each end of the 90 meter backbone achieves the maximum channel length of 100 meters.

Lower-speed data systems such as IBM 3270, IBM System 36, 38, AS/400 and asynchronous (RS232, RS422, RS423, etc.) can operate over UTP (or STP) for considerably longer distances, typically from several hundred feet to more than 1000 feet. The actual distances possible depend on the type of system, data speed, and the manufacturer's specifications for the system electronics and the associated components used (e.g., baluns, adapters, line drivers, etc.). Current state-of-the-art distribution facilities usually include a combination of both copper and fiber optic cables in the backbone.

Other design requirements for backbone cabling include:

- The cable plant should be based on a star topology (hub and spoke).
- The cable plant must contain no more than two hierarchical levels of cross-connects.
- Bridge taps are not allowed.
- Main and intermediate cross-connect jumper or patch cord lengths should not exceed 20 meters (66 feet).
- Avoid installing in areas where sources of high levels of EMI/RFI may exist.
- Grounding should meet the requirements as defined in TIA/EIA-607.

4. Telecommunications Closet

Telecommunications closets are the areas within a building that house the telecommunications cabling system equipment. This includes the mechanical terminations and/or cross-connect for the horizontal and backbone cabling system. As with equipment rooms, the design specifics for telecommunications closets are also defined in the TIA/EIA 569-A standard.

5. Horizontal Cabling

The specified topology for horizontal cabling systems is a *star*. Horizontal cabling extends from the work area outlet to the telecommunications closet and consists of:

- Horizontal cabling
- Telecommunications outlet
- Cable terminations
- Cross-connections

Three media types are recognized as options for horizontal premise cabling, each extending a maximum distance of 90 meters:

- 4-pair, 100-Ohm UTP cable (24 AWG solid conductors)
- 2-pair, 150-Ohm STP cables
- 2-fiber, 62.5/125-μm optical cable

Note: At this time, 50-Ohm coaxial cable is a recognized media type. It is not, however, recommended for new cabling installations and is expected to be removed from the next revision of the standard.

In addition to the 90 meters of horizontal premise cable, a total of 10 meters is allowed for work area and telecommunications closet patch and jumper cables. The original standard permitted this 10 meters to be divided a number of ways between work area and telecommunications closet, such as 7 meters for the telecom closet and 3 meters for the work area. The standard now recommends the length be distributed evenly, with 5 meters for each. This measurement is intended to prevent cable channels from accidentally exceeding the 100-meter limit (see Figure 12.8).

The horizontal cabling specifications also define the eight-position modular jack (RJ-45) and the ANSI/TIA/EIA 568-A pin assignment configurations for both T568-A and T568-B.

6. Work Area

Work area components extend from the telecommunications outlet to the end-station equipment. Work area wiring is designed to be relatively simple to interconnect so that moves, adds, and changes are easily managed. Work area components include:

- Station equipment: computers, data terminals, telephones, etc.
- Patch cables: modular cords, PC adapter cables, fiber jumpers, etc.
- Adapters (baluns, etc.): *must be external to telecommunications outlet!*

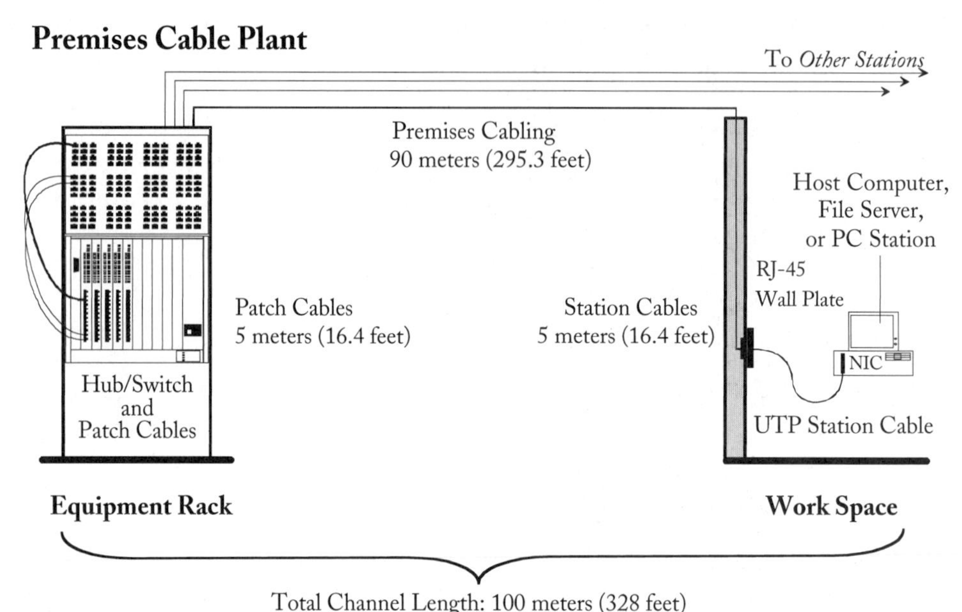

Figure 12.8 *ANSI TIA/EIA 568-A Horizontal Cabling*

Media and Connecting Hardware Performance Specifications

Also part of ANSI/TIA/EIA 568-A are the specifications for the various categories of UTP cable used in the construction of the Structured Cabling System. The current categories include Cat3 and Cat5E. Categories 1 and 2 were discontinued some time ago. Category 4 was superceded by Category 5, and Category 5 has been replaced by Category 5E (enhanced). Several standards bodies have published specifications for Category 6 media, but Category 7 is still a work in progress. While each category of UTP shares a few common characteristics, the key difference is in their spectral bandwidth. Wider bandwidth means greater carrying capacity—i.e., the potential to transmit more bits per second. Table 12.3 provides a comparison of past, present, and potentially future UTP categories.

In addition, remove only as much of the end of the cable jacket or sheath as necessary to terminate on connecting hardware (about 2–3 inches maximum). Run two sets of four-pair UTP to each work area. One set must be Cat5E, the other Cat3 or better. Do not tightly cinch bundles of UTP cables with cable ties, and space the cable ties on irregular intervals. Screw terminals are not permitted, nor are bridged taps, bridge clips, or splices. Minimize the use of cross-connects to keep noise to a minimum.

The bulk of the cable specifications place limits on acceptable levels of signal loss and noise. Not so long ago, everyone thought "wire was wire"—that any cable was just about as good as any other—but today we know different, and it seems there is always

TABLE 12.3 ANSI/TIA/EIA 568-A UTP CATEGORIES

Category	*Media*	*Spectral B/W*	*Channel Length*	*Applications*
Cat3	UTP	16 MHz	100 meters	10Base-T, 4 Mb token ring
Cat4	UTP	20 MHz	100 meters	Obsolete, superceded by Cat5
Cat5	UTP	100 MHz	100 meters	Obsolete, superceded by Cat5E
Cat5E	UTP	100 MHz	100 meters	100Base-TX, ATM, 1000Base-T
Cat6	UTP	250 MHz	100 meters	TIA Draft 1000Base-TX, n/a
Cat6E*	UTP	250 MHz	100 meters	n/a
Cat7*	STP	600 MHz	100 meters	n/a
Cat8**	?	?	?	n/a

*Not yet standards in this country.

**Not yet a draft proposal anywhere.

something new to learn about measuring its performance characteristics. Performance measurements that have been around since 10 Mbps Classic Ethernet include the characteristic impedance of the wire, attenuation, near-end crosstalk (NEXT), Multiple Disturber NEXT (MDNEXT), and impulse noise. With the advent of 100 Mbps Fast Ethernet and 1000 Mbps Gigabit Ethernet came Category 5 and Category 5E, and with them, more stringent cable plant performance measurements. New cable plants using these new media require performance measurements for Power Sum NEXT (PSNEXT), worst pair-to-pair ELFEXT, propagation delay, delay skew, and return loss.

Major differences between Anixter Levels and ANSI/TIA/EIA standards are:

- Standards provide performance requirements for the components and the link without specifying or identifying a matched system. Anixter tests components and channels and can explain how the components will work together as a system.
- Standards allow for impedance smoothing (averaging); Anixter does not.
- Anixter's patch cord specifications include both return loss and NEXT.
- Anixter will not accept spliced cable or non-virgin materials.
- Anixter requires power sum testing for connecting hardware (as well as for cable and channel), while the standards do not (i.e. all pairs energized).

ANSI/TIA/EIA 569-A Design Considerations

The ANSI/TIA/EIA 569-A Design Considerations standards are intended to provide for a generic, structured cabling plant, capable of running any voice or data application foreseeable in the next ten to fifteen years. The reality of cable plant technology is that the software, hardware and communications equipment that use it have much shorter lifespans of one to five years (due to obsolescence). This means that as applications and networking equipment are upgraded, the cable plant in most cases must also be upgraded to accommodate the new technology.

ANSI/TIA/EIA 606 Administration

As the name implies, the ANSI/TIA/EIA 606 Administration standard for the Telecommunications Infrastructure of Commercial Buildings involves lots of documentation—specifically, administrative records, reports, drawings, and work orders. This standard also defines unique *Identifiers* for each space, pathway, cable termination point, and ground. Identifiers are numbers that can be encoded in such a way as to provide supplemental information.

ANSI/TIA/EIA 607 Grounding and Bonding

ANSI/TIA/EIA 607 Commercial Building Grounding and Bonding Requirements for Telecommunications addresses the need for a uniform grounding and bonding infrastructure in commercial buildings. With the divestiture of AT&T in 1984, the end-user became responsible for all voice and data premises cabling. Technological advancements in voice and data communications, and the convergence of voice and data communications, has led to increasingly complex systems owned and maintained by the end-user. These systems require a reliable electrical ground-reference potential. Grounding by attachment to the nearest piece of iron pipe is no longer satisfactory to provide ground-reference for sophisticated active electronics systems.

CHAPTER SUMMARY

The wide world of wiring is a dynamic and ever-evolving place. There is tremendous variety in the available network media, both copper and glass. All cabling is not created equal, and some is not even interoperable with comparable product from other manufacturers. Careful attention must be paid to the electrical characteristics of any UTP cable plant installation, and even then the standards for gauging those characteristics are lacking in several areas. Cable specifications represent just one facet of structured cabling systems. ANSI/TIA/EIA has developed several standards that define not only the physical media but the design, administration, and even the proper grounding of a generic structured cabling system. These standards are required for any modern cable plant installation.

And although a certain type of cable may appear to have some advantages over another, that does not mean it will succeed in the long term. Even though it can support a greater bandwidth than UTP, coaxial cable is no longer used for LANs. Fiber has great potential, and can support the greatest bandwidth over the greatest distances, but the cost of the end electronics is still prohibitively high for wide-scale deployment. For distances of 100 meters or less, UTP can support data rates comparable to that of fiber. Fiber shines when the requirement is for distances in excess of 100 meters, noisy environments, and inter-building connections.

REVIEW QUESTIONS

1. Identify three types of coaxial cable, their impedance, and their applications.
2. Identify three types of STP cable, their impedance, and their applications.
3. Identify three types of UTP cable, their impedance, and their applications.
4. Identify three types of optical fiber and the wavelengths supported.
5. Define the pin assignments for T568-A and T568-B.
6. What is the maximum length of a Category 5E channel?
7. What does ANSI/TIA/EIA 568-A define?
8. Define the following acronyms:

AWG	ScTP
RG	DIW
PDS	USOC
PVC	Xtalk
NEXT	PSNEXT

9. What is plenum-rated cable?
10. What is a structured cabling system?

IP Addressing and Subnetting

OBJECTIVES

After completing this chapter, you should be able to:

- Identify the different versions of IP in use today.
- Describe the characteristics of Class A, B, and C addressing.
- Describe the differences between public and private IP addresses.
- Identify broadcast, network, and host addresses.
- Identify subnet masks and how to use them.
- Identify and describe the default gateway.
- Describe Classless Inter-Domain Routing.
- Describe Dynamic Host Configuration Protocol.
- Identify the three private IP address spaces.
- Identify the Loopback address.

INTRODUCTION

In this final chapter you will learn about classful and classless Internet Protocol addressing, as well as how to create and identify subnets and supernets. In addition, you will learn how to configure host systems using static and dynamic IP addressing.

IPv4 ADDRESSING

The communications protocol used in the public Internet as well as most private intranets today is known as the Internet Protocol, or simply IP. The current standard for IP is version 4, or IPv4, which uses an addressing system based on a 32-bit (4 octet) number. The popularity of the Internet has caused concern that the current IPv4 address space will be inadequate. This has prompted the development of a new and improved Internet Protocol called IP version 6, or IPv6 (briefly known as IP the Next Generation—IPng). IPv6 uses 128-bit (16 octet) addressing, but is not as widely deployed as IP version 4. In addition to a vastly larger address space, IPv6 provides other improvements over IPv4. However, interest in migrating to this new version is more keenly felt overseas in emerging technological societies such as China than here in the United States, which controls about 75% of the current IPv4 address space. The remainder of this text will focus on the ubiquitous IPv4.

Two methods of IP addressing have been developed: *classful* and *classless*. To understand and manipulate IP addressing requires a basic foundation in binary (i.e., counting through 24 bits and some Boolean math). Binary refers to the Base2 number system—counting with ones and zeros. Most of our fellow humans tend to prefer decimal, or Base10, which uses the numbers 0 through 9. Computer systems and programs have at various times relied on octal (Base8, which uses the numbers 0 through 7) and, more

TABLE 13.1 THE NUMBERS FROM DECIMAL 0 THROUGH 15 REPRESENTED IN DECIMAL, BINARY, OCTAL, AND HEXADECIMAL FORM ARE LISTED BELOW

Decimal (Base10)	*Binary (Base2)*	*Hexadecimal (Base16)*	*Octal (Base8)*
0	0000	00	00
1	0001	01	01
2	0010	02	02
3	0011	03	03
4	0100	04	04
5	0101	05	05
6	0110	06	06
7	0111	07	07
8	1000	08	10
9	1001	09	11
10	1010	0A	12
11	1011	0B	13
12	1100	0C	14
13	1101	0D	15
14	1110	0E	16
15	1111	0F	17

commonly, hexadecimal (Base16, which uses the numbers 0 through 9 plus A, B, C, D, E, and F). Table 13.1 shows the first sixteen integers for four different number systems: decimal (Base10), binary (Base2), octal (Base8), and hexadecimal (Base16).

Most IPv4 host addressing systems are configured and represented in decimal, or more specifically, dotted decimal notation: four octets separated by periods (255.255.255.255). To represent a single octet (or 8-bit byte) in this manner requires the decimal numbers 0 through 255. The highest number possible in any of the four octets is decimal 255 (which equates to 11111111 in binary, or FF in hexadecimal).

Before attempting to learn classless addressing (IETF rfc1519, rfc1918 and CIDR), it is perhaps best to first master classful addressing. The earlier and more common of the two is classful addressing, which defines five different number ranges identified as Classes A, B, C, D, and E. Three of these address classes are used to define IP networks and host addresses: A, B, and C. Each address class defines a different range of numbers and specifies a different number of network addresses. Each network class also supports a different number of host addresses per network. Network Classes D and E are reserved for multicast and experimental applications respectively, and as such are not essential to this tutorial.

IP ADDRESS CLASSES

Each class of IP address supports a range of network addresses, each of which supports a number of host addresses. Table 13.2 identifies all permissible IPv4 addresses.

An internetwork consists of multiple networks joined by routers. Internetworks may use publicly registered IP addresses, or private IP addresses as defined in IETF RFC 1918. Publicly registered IP addresses are required for direct access to the Internet, whereas private IP addresses may be used within any internetwork other than the Internet. Private internetworks require gateways to connect to the public Internet. Such gateways translate the private IP addresses used within the private internetwork into public IP addresses for transmission across the public Internet.

If using publicly registered IP addresses, the addressing scheme will be limited by the constraints of that address class. If using private IP addresses, a Class A, B, or C address can be chosen. The Class A network, 10.0.0.0, is the only Class A network reserved for private use and provides the most flexibility. Most SOHO networking products are configured from the factory to use Class C network 192.168.1.0 or 192.168.0.0. The ranges of IP address space shown in Table 13.3 are reserved for private networking.

Private IP addresses are often referred to as "invalid" or "non-routable" IP addresses. In addition to generating confusion, these descriptions are simply wrong. It is true that private IP addresses are prohibited from being routed over the public Internet; however, private IP addresses are absolutely valid for use in private internetworks, and they are certainly routable. To differentiate between the two types of IP addresses, refer to them simply as *private* and *public*.

TABLE 13.2 PERMISSIBLE IPv4 ADDRESSES

***Class A** supports a total of 126 networks and 16,777,214 host addresses per network.*	
Reserved:	0.0.0.0 ("This Host", and used to define the Default Route)
Network Range:	1.0.0.0 – 126.0.0.0
Host Range:	0.0.0.1 – 0.255.255.254
Reserved:	127.0.0.0 (Loopback, usually 127.0.0.1)
***Class B** supports a total of 16,382 (16k) networks and 65, 534 (64k) host addresses per network.*	
Reserved:	128.0.0.0
Network Range:	128.0.0.0 – 191.255.0.0
Host Range:	0.0.0.1 – 0.0.255.254
Reserved:	191.255.0.0
***Class C** supports a total of 2,097,150 networks and 254 host addresses per network.*	
Reserved:	192.0.0.0
Network Range:	192.0.0.0 – 223.255.255.0
Host Range:	0.0.0.1 – 0.0.0.254
Reserved:	223.255.255.0
***Class D** supports multicast and experimental applications.*	
Network Range:	224.0.0.0 – 239.255.255.255
***Class E** is reserved for future and experimental applications.*	
Network Range:	240.0.0.0 – 247.255.255.255
Network Range:	248.0.0.0 – 255.255.255.254
Reserved:	255.255.255.255 (Referred to as a "Limited Broadcast")

TABLE 13.3 PRIVATE IP ADDRESS RANGES HAVE BEEN DEFINED FOR CLASS A, B, AND C NETWORKS

Class	***Network Range***	***Netmask***	***Host Range***
A	10.0.0.0	(10/8 prefix)	10.0.0.1 – 10.255.255.254
B	172.16.0.0 – 172.31.0.0	(172.16/12 prefix)	172.16.0.1 – 172.31.255.254
C	192.168.0.0 – 192.168.255.0	(192.168/16 prefix)	192.168.0.1 – 192.168.255.254

In addition, while some organizations may use an arbitrarily selected public IP network address in their own private networks, this practice is NOT recommended. Although it is possible to use any network address in a private network, problems will arise if and when a user attempts to communicate with a host in the domain that actually owns that network address. This is because the IP network address of the destination host exists in the tables of the local routers. Because the local routers consider that network to be local, all packets addressed to that same network will be routed locally, never reaching the intended destination host. Moral of the story: do not use public addresses that do not belong to you!

169.254.0.0 to 169.254.255.255 is another reserved address space. RFC 3330 defines its purpose using the extremely vague term, "Link Local." This is a feature incorporated into the TCP/IP stack of some hosts. If such a host is configured to obtain an IP address automatically from a DHCP server and no DHCP server is available, that host's TCP/IP stack may automatically assign itself an IP address from this range (Automatic Private IP Address).

To execute this process the TCP/IP stack randomly selects an IP address from this range. The TCP/IP stack then transmits a gratuitous ARP broadcast onto the network. If no replies to the ARP are received, no address conflict currently exists and the host may begin using that IP address. However, this process does not define a default gateway or DNS, and therefore limits its practical use to communications only between hosts within its own subnet.

In addition, some sources claim this IP address range is included as another private IP address range (as per RFC 1918), which it is not. And contrary to popular belief, although Microsoft does employ this feature in most of its operating systems, the company does NOT own the 169.254.0.0 address space. Its purpose and applications are defined in RFC 3330, and that address range is available for use by anyone for the purpose as described above.

SUBNETS

Subnetting is used to divide large address spaces into smaller, more manageable blocks. Each network interface on a given router represents a unique network or subnet. No two network interfaces on a given router may connect to the same IP subnet (without special considerations—e.g., router software designed to provide redundant connections). It is also important to note that all IP networks can be subnetted, whether using public or private Class A, B, or C addresses.

A router is not required in order to facilitate communications between hosts residing on the same subnet. However, a router absolutely is required any time packets must be forwarded from one subnet to another, or between hosts residing on two different networks or subnets.

A single router interface may support multiple IP networks or subnetworks. This is often referred to as secondary IP addressing. While not considered good strategic planning, secondary IP addressing may be used to increase the total number of host addresses supported on a single physical network. For instance, adding a second Class-C network to a router interface increases the total number of host addresses supported from 254 to 508. However, hosts residing on those different subnets, although they may be attached to the same physical network, cannot "see" each other directly. All communications between subnets must pass through the router. This means that each IP packet transmitted between two hosts residing on the same physical network but in separate IP subnets must be transmitted across the physical network twice; from the source host on the first subnet over the LAN to the router, and from the router back over the same LAN to the destination host on the second subnet. Obviously, doubling the amount of network traffic by design is not a good idea, so try to avoid it.

All host computers on a given LAN segment should be configured as part of the same subnet defined by the router interface to which that segment is attached. Every device with an IP address is considered a host—including personal computers (PCs and Macs), Unix systems, mini-computers, mainframes, FEPs and controllers, network-capable personal information managers (PIMs), personal digital assistants (PDAs), IP-capable cell phones and other mobile data communications devices, LAN-attached printers, all router interfaces, and all SNMP-manageable devices on the network (such as hubs and switches).

In most internetwork designs, an IP network is divided into multiple subnetworks (or subnets). The optimum subnet architecture will depend on the needs and characteristics of the organization it will support, and the class of IP address used. Organizational characteristics include its overall size, the number of campuses, buildings per campus, floors per building, workgroups per building or floor, and computers per workgroup, including every device with an IP address.

You must therefore determine the total number of subnets required for the entire organization, and the total number of hosts required per subnet. This will vary from organization to organization, and over time—so get out your crystal ball.

An additional consideration often overlooked is organizational growth. Take time to discuss future organizational needs with everyone affected. A miscalculation in the addressing scheme of your internetwork design in the beginning of the project can be a daunting task to correct once implemented.

A common misconception is that subnetting provides more host addresses, when in fact the reverse is true. The more subnets defined for a given network, the fewer hosts each subnet will support. In addition, each subnet wastes two potential host addresses—the all-ones directed broadcast address, and the all-zeros network address.

A good rule of thumb—at least, a place from which to start—is to limit the number of hosts per subnet to about 120. This provides for a reasonably sized workgroup of users and shared resources and can be accommodated with a 7-bit host field. When workgroups get much larger than this, they tend to become unwieldy, difficult to manage and easily saturated with network traffic. Keep in mind that this is just a recommendation, not a law. This recommendation will not apply to all situations. And a simpler method is to subnet on the octet boundary (i.e., use a subnet mask of 8, 16, or 24 bits). Several examples are provided later in this chapter.

SUBNET MASK

A subnet mask is used to define the subnet field within the host portion of the IP address. In classful IP addressing, you typically do not alter or change the network portion of the address once a network address is assigned. However, you may customize the use of the host bits to define an appropriate subnet field. The size of the subnet field defines the total number of subnets available, as well as the total number of hosts per subnet.

Several rules and restrictions have been imposed on IP addressing. The host portion of an IP address may never consist of all ones or all zeros. Routers use the all-zeros host address to represent an entire subnet, and a host address of all-ones is the local broadcast address for that subnet. Technically referred to as a *directed broadcast*, each subnet has its own directed broadcast address—the all-ones host address of each subnet. Every IP network or subnet addressing scheme defines a range of addresses that may be allocated to hosts. Bounding each range are two addresses reserved for special purposes: the network address and the directed broadcast address. Here are two examples of IP network address ranges.

The following Class C network example assumes a subnet mask of 255.255.255.0:

192.168.1.0	Reserved to represent the entire network
192.168.1.1	The first host address for network 192.168.1.0
192.168.1.254	The last host address for network 192.168.1.0
192.168.1.255	Reserved as the directed broadcast address for network 192.168.1.0

The host address 192.168.1.12 represents host 12 on network 192.168.1.

The following Class B network example assumes a subnet mask of 255.255.0.0:

172.16.0.0	Reserved to represent the entire network
172.16.0.1	The first host address for network 172.16.0.0
172.16.255.254	The last host address for network 172.16.0.0
172.16.255.255	Reserved as the directed broadcast address for network 172.16.0.0

The host address 172.16.5.12 represents host 5.12 on network 172.16.

In a given network, the subnet field of an IP address may consist of almost any number of the available host bits, even a single bit. Until a few years ago, a subnet value of all ones or all zeros was prohibited. Because a subnet field of just one bit provides only two subnet values, subnet 0 and subnet 1, one-bit subnet fields were unusable. This is no longer the case. The old rule has since been rescinded, thereby permitting the use of subnet values consisting of all ones or all zeros, including one-bit subnet fields. Even so, most network administrators still avoid using these networks. And as mentioned above, no host may be assigned an IP address consisting of all ones or all zeros in the host field (the network address or directed broadcast address).

A subnet field consists of one or more consecutive bits immediately following the network field. You may not skip bits within the subnet field (i.e., 11000111). The only legal subnet mask values are shown in Table 13.4. (Commit this table of IP mask values to memory—it will serve you well.)

It is worth noting that the term *subnet mask* may be perceived as a bit of a misnomer, since it actually appears to mask the host bits rather than the subnet bits. The ultimate purpose of the subnet mask is to define (and subsequently reveal) the network and subnet portion of an IP address. Routers forward IP packets based on the destination IP address found in the header of each packet. Routers make their forwarding decisions based on the network and subnet portion of this address, not the host portion. Routers use the subnet mask to determine which bits of the address define the network and subnet, and which bits define the hosts on that network/subnet.

The routing of packets on the public Internet is not affected by the internal subnetting scheme of any private internetwork (using public or private addresses). Routing information generated by routers inside a private internetwork is not permitted on the

TABLE 13.4 SUBNET MASK VALUES

Binary Mask	*Decimal Mask*	*Subnet Bits*	*Host Bits*
00000000	0	0	8
10000000	128	1	7
11000000	192	2	6
11100000	224	3	5
11110000	240	4	4
11111000	248	5	3
11111100	252	6	2
11111110	254	7	1
11111111	255	8	0

public Internet. Therefore, any routing technology may be deployed inside private internetworks regardless of the routing technology supported by the ISP.

In the examples below, "N" defines the network bits and "H" defines the host bits. You may use the host bits any way you wish—e.g., to define subnets and to identify individual hosts in those subnets. Dotted decimal notation (Base10) is used to refer to the four octets of the IPv4 address.

Classful IP Addressing

Class A network addresses occupy the first octet:	N.H.H.H
Class B network addresses occupy the first two octets:	N.N.H.H
Class C network addresses occupy the first three octets:	N.N.N.H
Class D and E network addresses occupy all four octets:	N.N.N.N

Each address class may be identified by its leading bits (n = network bits & h = host bits):

Class A	0nnnnnnn.hhhhhhhh.hhhhhhhh.hhhhhhhh
Class B	10nnnnnn.nnnnnnnn.hhhhhhhh.hhhhhhhh
Class C	110nnnnn.nnnnnnnn.nnnnnnnn.hhhhhhhh
Class D	1110nnnn.nnnnnnnn.nnnnnnnn.nnnnnnnn
Class E	11110nnn.nnnnnnnn.nnnnnnnn.nnnnnnnn

A default mask is associated with each class of address. The default mask places all ones in the network portion and zeros in the host portion of the address:

Class A (binary):	0nnnnnnn.hhhhhhhh.hhhhhhhh.hhhhhhhh
Default mask (binary):	11111111.00000000.00000000.00000000
Default mask (decimal):	255.0.0.0
Class B (binary):	10nnnnnn.nnnnnnnn.hhhhhhhh.hhhhhhhh
Default mask (binary):	11111111.11111111.00000000.00000000
Default mask (decimal):	255.255.0.0
Class C (binary):	110nnnnn.nnnnnnnn.nnnnnnnn.hhhhhhhh
Default mask (binary):	11111111.11111111.11111111.00000000
Default mask (decimal):	255.255.255.0
Class D (binary):	1110nnnn.nnnnnnnn.nnnnnnnn.nnnnnnnn
Class E (binary):	11110nnn.nnnnnnnn.nnnnnnnn.nnnnnnnn
Limited broadcast (binary):	11111111.11111111.11111111.11111111

IP SUBNETTING EXAMPLES

The following examples illustrate some commonly used subnetting schemes for Class A, B, and C IP addresses. Example 1 shows a Class A network using a 16-bit subnet field to provide 65,536 subnets with support for 254 hosts per subnet. Example 2 shows a Class B network using an 8-bit subnet field to provide 256 subnets with support for 254 hosts per subnet.

Note that while both example 1 and example 2 use the same subnet mask, the resulting number of subnets differs because each IP address Class provides a different number of bits for subnet and host addressing. Class A networks are defined using the first 8 bits; leaving 24 bits for subnet and host addressing. Class B networks are defined using the first 16 bits; leaving just 16 bits for subnet and host addressing. Since both examples use the same subnet mask the number of subnets varies but the number of hosts remains the same.

Since there are more Class C addresses than Class A or Class B combined, several Class C examples are provided. Example 3 shows a Class C network using a 4-bit subnet field to provide 16 subnets with support for 14 hosts per subnet. Example 4 shows another Class C network using a 5-bit subnet field to provide 32 subnets with support for 6 hosts per subnet. And example 5 shows one more Class C network using a 6-bit subnet field to provide 64 subnets with support for just 2 hosts per subnet.

Note that while these three examples all use a Class C address, each defines a different subnet mask. The resulting number of subnets and hosts varies because each Class C network example (examples 3, 4, and 5) provides an increasing number of bits for subnet addressing, and therefore leaves fewer bits for host addressing.

Example 6 provides a variation on Class A subnetting using a 14-bit subnet field to provide 16,384 subnets with support for 1022 hosts per subnet. Such an addressing scheme may be used to satisfy a requirement to accommodate a large number of hosts in each subnet. Note that by shifting two bits from the subnet field to the host field, the number of possible hosts has been increased by (approximately) a factor of four and the number of possible subnets has been reduced by (exactly) a factor of four. This is due to the all-ones and all-zeros rules for subnet and host addresses.

LEGEND: **Bold** = Network field, Regular = Subnet field, *Italic* = Host field

Example 1:

Class A Network: 10.0.0.0
Default Mask: 255.0.0.0
Subnet Mask Example: 255.255.255.0
(Default Mask in binary): 11111111.*00000000.00000000.00000000*
(Subnet Mask in binary): **11111111**.11111111.11111111.*00000000*

The 1st octet defines the **network field.**

The subnet mask indicates a 16-bit subnet field using the 2nd and 3rd octets in their entirety (65,536 subnets).

The remaining 8 bits of the 4th octet define the *host field*, providing 254 hosts per subnet.

The total number of hosts supported in this scheme is 65,536 × 254 (or 16,646,144 hosts).

With no subnetting (not recommended!), you could have 16,777,214 hosts.

Example 2:

Class B Network:	172.16.0.0
Default Mask:	255.255.0.0
Subnet Mask Example:	255.255.255.0
(Default Mask in binary):	11111111.11111111.*00000000.00000000*
(Subnet Mask in binary):	**11111111.11111111**.11111111.*00000000*

The 1st and 2nd octets define the **network field**.

The subnet mask indicates an 8-bit subnet field using the 3rd octet (256 subnets).

The remaining 8 bits of the 4th octet define the *host field*, providing 254 hosts per subnet.

The total number of hosts supported in this scheme is 256 × 254 (or 65,024 hosts).

With no subnetting (again not recommended!), you could have 65,534 hosts.

Example 3:

Class C Network:	192.168.1.0
Default Mask:	255.255.255.0
Subnet Mask Example:	255.255.255.240
(Default Mask in binary):	11111111.11111111.11111111.*00000000*
(Subnet Mask in binary):	**11111111.11111111.11111111**.1111*0000*

The 1st, 2nd, and 3rd octets define the **network field**.

The subnet mask indicates a 4-bit subnet field using half of the 4th octet (16 subnets).

The remaining 4 bits of the 4th octet define the *host field*, providing 14 hosts per subnet.

The total number of hosts supported in this scheme is 16 × 14 (or 224 hosts).

With no subnetting, you could have 254 hosts.

This configuration is commonly used by ISPs to support 16 small business customers using a single Class C address.

Example 4:

Class C Network:	192.168.1.0
Default Mask:	255.255.255.0
Subnet Mask Example:	255.255.255.248
(Default Mask in binary):	11111111.11111111.11111111.*00000000*
(Subnet Mask in binary):	**11111111.11111111.11111111**.11111*000*

The 1st, 2nd & 3rd octets again define the **network field**.

The subnet mask indicates a 5-bit subnet field using more of the 4th octet (32 subnets).

The remaining 3 bits of the 4th octet define the *host field*, providing 6 hosts per subnet.

The total number of hosts supported in this scheme is 32 × 6 (or 192 hosts).

With no subnetting (which is often used), you could have 254 hosts. It is very common to use no subnetting with Class C addresses.

This configuration is also commonly used by ISPs to support 32 small business customers with one Class C address.

Example 5:

Class C Network:	192.168.1.0
Default Mask:	255.255.255.0
Subnet Mask Example:	255.255.255.252
(Default Mask in binary):	11111111.11111111.11111111.*00000000*
(Subnet Mask in binary):	**11111111.11111111.11111111**.111111*00*

The 1st, 2nd, and 3rd octets define the **network field**.

The subnet mask indicates a 6-bit subnet field using even more of the 4th octet (64 subnets).

The remaining 2 bits of the 4th octet define the *host field*, providing just 2 hosts per subnet. (01 and 10 are valid host addresses while 00 and 11 are reserved.)

The total number of hosts supported in this scheme is 64 × 2 (or 128 hosts).

With no subnetting, you could again have 254 hosts.

This odd configuration is commonly used to address point-to-point WAN circuits. (Such circuits support only two hosts—one router at each end.)

Example 6:

Class A Network:	56.0.0.0
Default Mask:	255.0.0.0

Subnet Mask Example: 255.255.252.0
(Default Mask in binary): 11111111.*00000000.00000000.00000000*
(Subnet Mask in binary): **11111111**.11111111.111111*00.00000000*

The 1st octet defines the **network field**.

The subnet mask indicates a 14-bit subnet field (16,384 subnets).

The remaining 10 bits define the *host field*, providing 1022 hosts per subnet.

The total number of hosts supported in this scheme is 16,384 × 1022 (or 16,744,448 hosts).

With no subnetting (still not recommended!), you could have 16,777,214 hosts.

The results of Example 6 may also be expressed as follows:

Total Subnets possible:
2^{14} 16,384 (0 – 16,383)

Total Hosts per Subnet possible:
2^{10} (–2) 1,022 (1 – 1,022)

Host Address Range:

Subnet #	**Network**	**First Host**	**Last Host**	**Directed Broadcast**
0	56.0.0.0	56.0.0.1	56.0.3.254	56.0.3.255
1	56.0.4.0	56.0.4.1	56.0.7.254	56.0.7.255
2	56.0.8.0	56.0.8.1	56.0.11.254	56.0.11.255
3	56.0.12.0	56.0.12.1	56.0.15.254	56.0.15.255
4	56.0.16.0	56.0.16.1	56.0.19.254	56.0.19.255
5	56.0.20.0	56.0.20.1	56.0.23.254	56.0.23.255
.				
.				
.				
16,379	56.255.236.0	56.255.236.1	56.255.239.254	56.255.239.255
16,380	56.255.240.0	56.255.240.1	56.255.243.254	56.255.243.255
16,381	56.255.244.0	56.255.244.1	56.255.247.254	56.255.247.255
16,382	56.255.248.0	56.255.248.1	56.255.251.254	56.255.251.255
16,383	56.255.252.0	56.255.252.1	56.255.255.254	56.255.255.255

Now to put all this information to use! Routers and hosts use the subnet mask to define the subnet bits of the IP address. This is accomplished by performing a Boolean *AND* function between the destination IP address in a packet and the established subnet mask on the router or host. Without going into excessive detail certain to send the reader into a coma, Boolean functions typically include *AND*, *OR*, and *XOR*, and rely on so-called *truth tables* such as the one below:

A		B	Result
0	*AND*	0	= 0
1	*AND*	0	= 0
0	*AND*	1	= 0
1	*AND*	1	= 1

For the purposes of determining the subnet field of an IP address, we are concerned only with the Boolean *AND* function. To apply this concept to IP addresses and subnet masks, first convert the addresses to binary and then apply the above truth table to the addresses as follows:

HostAddress:	192.168.1.39	11000000.10101000.00000001.00100111
AND		
Subnet Mask:	255.255.255.240	11111111.11111111.11111111.11110000
RESULT 5	***192.168.1.32***	***11000000.10101000.00000001.00100000***

Translation: The result of the Host Address *AND* the Subnet Mask is the network/subnet address. The value of the host bits has been omitted. The host bits are not needed until the packet reaches the destination network or subnet. The network/subnet portion of an IP address is all the information required by routers or hosts to make forwarding decisions. If the IP address in a given packet is not local to that router or host, the packet will be forwarded to the next router hop, or in the case of a host, the packet will be forwarded to the default gateway.

SUPERNETTING WITH CIDR

And now for something completely different... Whereas classful IP addressing supports only subnetting, Classless Inter-Domain Routing supports supernetting. CIDR (pronounced "cider") is a relatively recent method of IP addressing designed to reduce the size of routing tables and provide greater address allocation flexibility. CIDR ignores network class conventions and provides the ability to aggregate many smaller, consecutively numbered networks into one large network address space.

To understand the benefit of doing this, consider the options faced by an organization that has outgrown its Class C network. One option is to switch to a Class B network, which would require the organization to reconfigure every host in the internetwork with the new addressing information. Another option would be to obtain a second Class C network, but this imposes certain limitations. Because the second Class C network is an entirely different network, the new address space cannot be used to expand the existing networks; the new address space can only be used to create new subnets.

As explained above, Class B networks provide 65,534 host addresses, and Class C networks provide 254 host addresses (even fewer if that network is subnetted). There are many organizations that have more than 254 hosts, but far fewer than 65,534. To provide IP addresses for all of their hosts, these organizations would have to obtain either multiple Class C networks, or a single Class B network. Still larger organizations may require multiple Class B networks, or one or more Class A networks.

Of course, this is a problem only if registered Internet addresses are required for each host. If the organization can use private IP addresses, then one or more private Class A, B, or C networks may be selected (or any combination thereof). The use of private IP addresses necessitates the use of Internet proxies or gateways that support Network Address Translation (NAT). Some applications do not work so well in NAT environments. Such applications may require the use of public IP addresses.

There are only 126 Class A networks and just 16,382 Class B networks available on the public Internet, and most are already in use. However, there are more than 2 million Class C networks. Clearly, this finite public address space must be allocated judiciously. Allocating a Class B network when several Class C networks would do may squander valuable public address space. Adhering to the old classful addressing conventions eventually became too restrictive, which is what prompted the IETF to develop CIDR and introduce the concept of supernetting.

Whereas subnetting uses some of the *host bits* of an IP network to divide one large address space into multiple smaller subnetworks, supernetting uses some of the *network bits* of an IP network to combine multiple smaller networks into one larger address space. Individually, each Class C network could support just 254 host addresses. The following example combines eight Class C networks to provide 2,046 host addresses into one large address space. Note the last three bits of the 3rd octet in each network address are consecutive:

Networks	3rd Octet in Binary
192.168.0.0	00000 000
192.168.1.0	00000 001
192.168.2.0	00000 010
192.168.3.0	00000 011
192.168.4.0	00000 100
192.168.5.0	00000 101
192.168.6.0	00000 110
192.168.7.0	00000 111

Instead of defining a subnet mask that creates a subnet field in the 4th octet of each Class C network previously mentioned, CIDR allows us to define a "supernet mask" called an IP prefix. An IP address with a CIDR prefix looks like a conventional IP address except that it ends with a slash followed by a number. It can be used to aggregate the eight IP networks in the example previously mentioned by defining an IP prefix of /21 (slash twenty-one). The result is called a CIDR block and is represented as follows: 192.168.0.0 /21. The number "21" refers to the amount of bits in the net mask. Another way to view this is in binary. The IP prefix equates to a netmask of 255.255.248.0.

Network Address	192.168.0.0 /21	11000000.10101000.00000000.00000000
Network Mask	255.255.248.0	11111111.11111111.11111000.00000000

Note that although this is a Class C network address, the CIDR prefix identifies eleven host bits in the network mask, providing a total of 2047 possible host addresses (0–2,046). The all-ones and all-zeros host addresses are still reserved, leaving a total of 2045 host addresses. Table 13.5 lists all possible CIDR prefixes, the equivalent decimal mask, and the number of available addresses. The last three columns identify the number of /24, /16, and /8 networks possible with each CIDR prefix.

TABLE 13.5 THE RANGE OF POSSIBLE CIDR PREFIXES, INCLUDING THE DECIMAL MASK AND NUMBER OF AVAILABLE ADDRESSES PROVIDED

CIDR Prefix	*Decimal Netmask*	*Number of Addresses*	*Number of /24 Networks*	*Number of /16 Networks*	*Number of /8 Networks*
/0	0.0.0.0	4,294,967,296	16,777,216	65,536	256
/1	128.0.0.0	2,147,483,648	8,388,608	32,768	128
/2	192.0.0.0	1,073,741,824	4,194,304	16,384	64
/3	224.0.0.0	536,870,912	2,097,152	8,192	32
/4	240.0.0.0	268,435,456	1,048,576	4,096	16
/5	248.0.0.0	134,217,728	524,288	2,048	8
/6	252.0.0.0	67,108,864	262,144	1,024	4
/7	254.0.0.0	33,554,432	131,072	512	2
/8	255.0.0.0	16,777,216	65,536	256	1
/9	255.128.0.0	8,388,608	32,768	128	
/10	255.192.0.0	4,194,304	16,384	64	
/11	255.224.0.0	2,097,152	8,192	32	
/12	255.240.0.0	1,048,576	4,096	16	

CIDR Prefix	Decimal Netmask	Number of Addresses	Number of /24 Networks	Number of /16 Networks	Number of /8 Networks
/13	255.248.0.0	524,288	2,048	8	
/14	255.252.0.0	262,144	1,024	4	
/15	255.254.0.0	131,072	512	2	
/16	255.255.0.0	65,536	256	1	
/17	255.255.128.0	32,768	128		
/18	255.255.192.0	16,384	64		
/19	255.255.224.0	8,192	32		
/20	255.255.240.0	4,096	16		
/21	255.255.248.0	2,048	8		
/22	255.255.252.0	1,024	4		
/23	255.255.254.0	512	2		
/24	255.255.255.0	256	1		
/25	255.255.255.128	128			
/26	255.255.255.192	64			
/27	255.255.255.224	32			
/28	255.255.255.240	16			
/29	255.255.255.248	8			
/30	255.255.255.252	4			
/31	255.255.255.254	2			
/32	255.255.255.255	1			

IP HOST CONFIGURATION

In IP parlance, the term *host* is not reserved just for servers, mini-computers, or mainframes. The term host refers to all network-attached devices with an IP address. All hosts require an IP address, and all addresses must be unique within a given network. An IP address is like a phone number. All phone numbers are unique and are used to identify every device in the telephone network with which you may communicate. Whereas telephone numbers are programmed in telephone switches, IP addresses are configured directly on every IP host.

Every IP host is usually configured with several other pieces of information in addition to its IP address. The minimum requirement to permit communication between any two hosts in the same IP network is for each host to have a different IP host address. The network and subnet portions of the IP addresses must be the same, but the host portion must be different. If the network has not been subnetted, use the default mask as the subnet mask. If the network has been subnetted, an appropriate subnet mask must also be defined.

To communicate with hosts on different IP networks or subnets requires a default gateway. The default gateway is a router that connects a given network or subnet to other networks or subnets. The default gateway for a given host is a router that is physically connected to that host's network. In other words, each host on a given subnet must know the IP address of the router that connects that subnet to the rest of the internetwork. That router is called the default gateway (or default router). When the location of a destination or target host is unknown, the packets are sent to the default gateway. Because the default gateway is a router, it contains routing tables that identify the best path to reach the destination host.

Hosts and routers make packet-forwarding decisions based on the 32-bit destination IP address contained in each packet. To simplify matters for the human operators, another form of host identification was developed—the fully qualified domain name (FQDN). An example of such a domain name is www.BuddyShipley.com. However, such names cannot be used to address IP packets. Before packets can be transmitted, these user-friendly names must be translated into the 32-bit numerical IP address corresponding to that IP host. The domain name system (DNS) was developed to provide the required translation service.

DNS is like an automated telephone white pages look-up service. If the fully qualified domain name is known, a DNS query can automatically provide the numerical IP address. DNS lookup is built into all TCP/IP protocol stacks. To use DNS requires defining one or more DNS server addresses on each host using the 32-bit numerical IP addresses of those DNS servers. The purpose in defining more than one DNS server is simply to increase reliability. If one DNS server fails, the TCP/IP protocol stack will try the second or third DNS server until it gets a response. More than three DNS server entries are usually considered overkill, because it is highly unlikely that two or three DNS servers would fail at the same time. It is more likely that some other network failure has disrupted communications to all hosts, including the DNS servers.

So in summary, each IP host must be configured with a unique IP address for identification, a subnet mask to determine the network and subnet on which it resides, a default gateway to communicate with hosts elsewhere across the internetwork, and a DNS server or two to support the use of user-friendly host names (without which users would be required to know the 32-bit IP address of every host with which they wished to communicate).

Common rules for IP addressing:

- Each logical network must have its own unique network address.
- All hosts in a given network must share the same network address.
- All hosts in a given network must have a unique host address.
- The host portion of the address may not consist of all ones or all zeros.
- Do not use IP addresses from the public address space unless specifically assigned.

Common requirements for all IP hosts:

IP Address	192.168.1.n	(Where n = any unused number from 2–254)
Subnet Mask	255.255.255.0	(Indicates this Class C network is not subnetted)
Default Gateway	192.168.1.1	(Typically uses the first or last host address)
DNS Server(s)	205.171.3.65	(Qwest)
	216.148.227.68	(ATT Broadband)
	204.127.202.4	(ATT Broadband)

Requirements unique to Microsoft networking:

WINS	192.168.1.x	(where x = the WINS Server computer)
NT Domains	192.168.1.y	(where y = the NT Domain Controller)
Active Directory	192.168.1.z	(where z = the AD Controller)

Most Microsoft networks use NetBIOS computer names to identify each computer. NetBIOS computer-naming conventions are not directly compatible with IP host-naming conventions. Each convention imposes different restrictions with regard to the use of special characters and spaces.

The term *WINS* refers to the Microsoft Windows Internet Naming Service, which runs in a Windows NT Server or Windows 2000/2003 Server. Just as DNS resolves IP host names to their numeric IP addresses, WINS resolves NetBIOS computer names to their numeric IP addresses. Although still in use, WINS has been superceded by other, more advanced name resolution systems referred to as *directory services*.

The first replacement for the WINS system was the Windows NT Primary and Backup Domain Controllers. Microsoft Active Directory (AD) and AD Controllers then replaced the NT Domain system. Most Microsoft Windows networks today rely on Active Directory rather than NT Domains or WINS.

DHCP

IP addresses may be assigned in one of two ways: statically or dynamically. Static IP addresses are manually configured on each host and remain the same until manually changed. IP addresses may also be dynamically assigned using the Dynamic Host Configuration Protocol (DHCP). In addition to the IP address, DHCP is also configured to supply the subnet mask, default gateway, DNS, and other optional IP addressing information.

DHCP is an application that runs on one or more servers in an internetwork. DHCP server applications are provided with several operating systems, such as Windows NT/2000 Server and Unix/Linux. In addition, most routers also include DHCP server capabilities, including inexpensive SOHO routers.

The TCP/IP stack of most operating systems provides an option to configure the host as a DHCP client. For Microsoft Windows systems, simply select the TCP/IP option to "obtain IP address automatically." Many host systems and other network-attached peripherals are configured to use DHCP by default from the factory.

Before DHCP existed, a function of Reverse Address Resolution Protocol (RARP) was used to dynamically allocate IP addresses. This system required a RARP server and a manually created table that defined all of the MAC addresses of any host that would ever request an IP address. This system was just as cumbersome as it sounds and never gained much popularity.

In contrast to the static table approach used by RARP, DHCP uses a concept of *scopes* and allocates IP addresses from dynamic pools. A scope defines all of the IP addressing properties of a single subnet, including ranges of IP addresses, the subnet mask, default gateway, domain name servers, WINS servers, and currently about 72 other optional parameters that are supported by some vendors. The IP address pool may consist of one or more ranges of addresses, or individual addresses. Some DHCP systems may be configured with one or more ranges of addresses, and then allow individual addresses to be excluded from inside those ranges. Different DHCP servers provide much more flexibility than others.

Because DHCP requests rely on MAC layer broadcasts, routers will not forward those requests—at least, not by default. Some routers can be configured to forward DHCP requests. This usually does not include the inexpensive SOHO variety. To configure a Cisco router to relay DHCP requests from a local subnet to a DHCP server located on a different subnet requires a special command, the syntax for which can appear quite cryptic to the uninitiated: ip forward-protocol udp bootpc.

DHCP simplifies management of IP addresses by automatically allocating addresses to new hosts as needed. Adding a new host is a matter of simply plugging it into the network. Moves and changes are also simplified, as DHCP automatically returns IP addresses that are no longer in use to the pool of available addresses. With DHCP, the daunting prospect of readdressing all of an organization's hosts is vastly simplified. In conclusion, DHCP saves time and money for IT departments.

CHAPTER SUMMARY

TCP/IP provides powerful and flexible addressing capabilities to accommodate diverse internetwork architectures. IP addressing schemes have evolved over the years, first providing the ability to segment large address spaces into smaller, more manageable subnetworks, and later to aggregate many smaller address spaces into a larger, more flexible supernet. To simplify the management and administration of IP addresses, powerful tools such as DHCP have been developed. All contemporary computer operating systems include basic support for TCP/IP, including network-attached peripherals such as printers, and the protocol's list of enhanced features continues to grow. As more applications are migrated to TCP/IP and the demands on the protocol increase, it becomes increasingly important for IT professionals to stay abreast of new developments in IP addressing, administration, and management. Good sources of information on networking and protocols include books such as this one, standards organizations, developers and manufacturers of networking products, and Internet websites such as www.BuddyShipley.com.

REVIEW QUESTIONS

1. Identify the two versions of IP in use today.
2. What is a host?
3. How many networks are provided by Class A, Class B, and Class C?
4. Define the difference between public and private IP addresses.
5. An IP address with all-ones in the host field denotes what?
6. List all of the valid subnet mask values in decimal.
7. What is a default gateway?
8. Define the following acronyms:

   CIDR SOHO

   DHCP PIM

   FQDN PDA

   DNS WINS

   RARP AD
9. List the three private IP address spaces.
10. What IP address is reserved for Loopback?

Electromagnetic Spectrum: Bandwidth and Frequency

For more information on the electromagnetic spectrum, see the following Web sites:

www.ntia.doc.gov/osmhome/allochrt.pdf

www.altair.org/specmap.html

www.hyperphysics.phy-astr.gsu.edu/hbase/ems1.html

www.lectureonline.cl.msu.edu/~mmp/applist/Spectrum/s.htm

www.science.msfc.nasa.gov/newhome/help/glossfig1.htm

www.its.bldrdoc.gov/fs-1037/images/frqcharc.gif

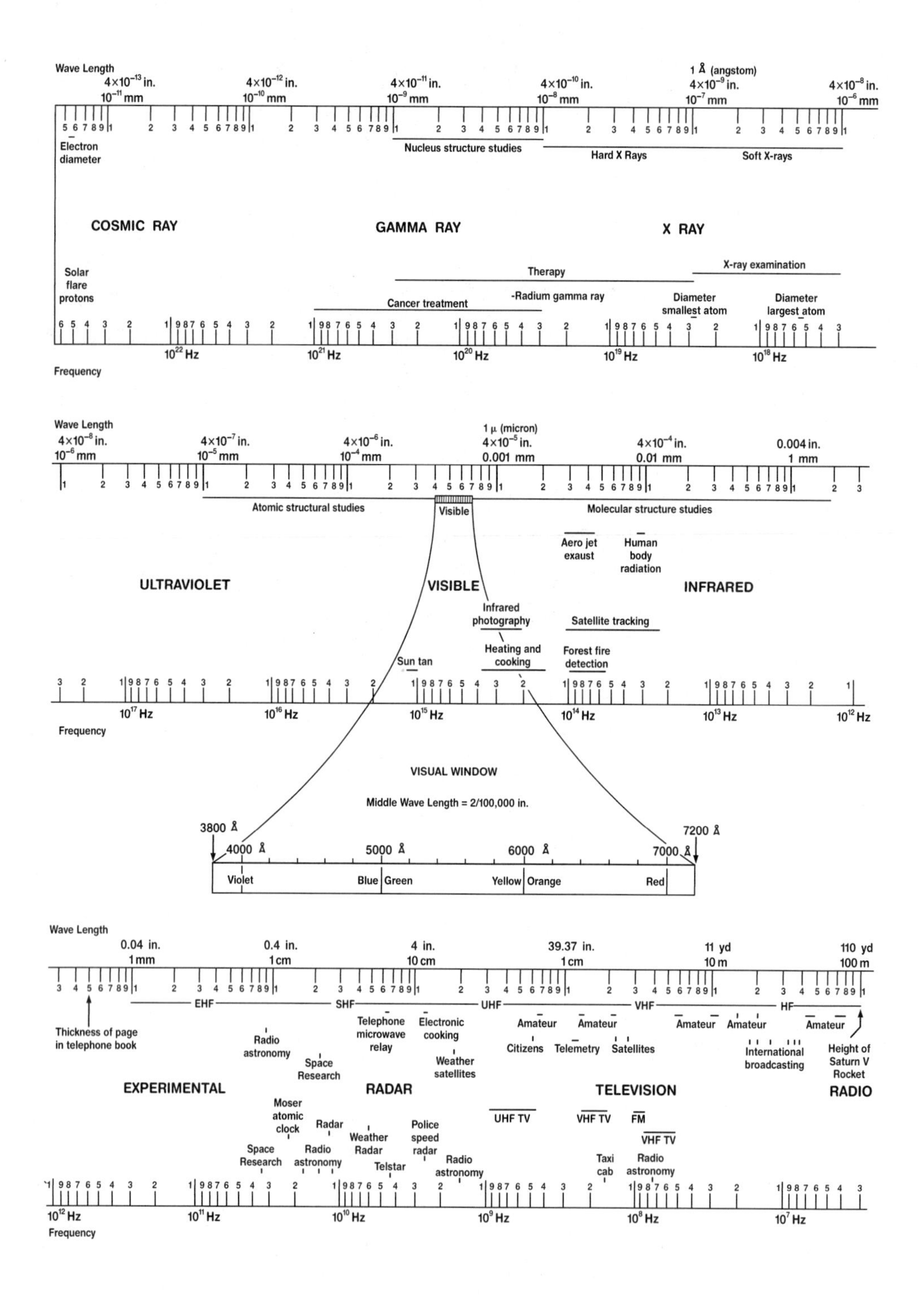

Wave Length
4×10⁻¹³ in. 10⁻¹¹ mm
4×10⁻¹² in. 10⁻¹⁰ mm
4×10⁻¹¹ in. 10⁻⁹ mm
4×10⁻¹⁰ in. 10⁻⁸ mm
1 Å (angstom) 4×10⁻⁹ in. 10⁻⁷ mm
4×10⁻⁸ in. 10⁻⁶ mm
Electron diameter
Nucleus structure studies
Hard X Rays
Soft X-rays
COSMIC RAY
GAMMA RAY
X RAY
Solar flare protons
Therapy
X-ray examination
Cancer treatment
-Radium gamma ray
Diameter smallest atom
Diameter largest atom
10²² Hz
10²¹ Hz
10²⁰ Hz
10¹⁹ Hz
10¹⁸ Hz
Frequency
Wave Length
4×10⁻⁸ in. 10⁻⁶ mm
4×10⁻⁷ in. 10⁻⁵ mm
4×10⁻⁶ in. 10⁻⁴ mm
1 μ (micron) 4×10⁻⁵ in. 0.001 mm
4×10⁻⁴ in. 0.01 mm
0.004 in. 1 mm
Atomic structural studies
Visible
Molecular structure studies
Aero jet exaust
Human body radiation
ULTRAVIOLET
VISIBLE
INFRARED
Infrared photography
Satellite tracking
Heating and cooking
Forest fire detection
Sun tan
10¹⁷ Hz
10¹⁶ Hz
10¹⁵ Hz
10¹⁴ Hz
10¹³ Hz
10¹² Hz
Frequency
VISUAL WINDOW
Middle Wave Length = 2/100,000 in.
3800 Å
7200 Å
4000 Å
5000 Å
6000 Å
7000 Å
Violet
Blue
Green
Yellow
Orange
Red
Wave Length
0.04 in. 1 mm
0.4 in. 1 cm
4 in. 10 cm
39.37 in. 1 cm
11 yd 10 m
110 yd 100 m
EHF
SHF
UHF
VHF
HF
Thickness of page in telephone book
Radio astronomy
Space Research
Telephone microwave relay
Electronic cooking
Weather satellites
Amateur
Amateur
Citizens
Telemetry
Satellites
Amateur
Amateur
Amateur
International broadcasting
Height of Saturn V Rocket
EXPERIMENTAL
RADAR
TELEVISION
RADIO
Moser atomic clock
Radar
Weather Radar
Police speed radar
UHF TV
VHF TV
FM
VHF TV
Space Research
Radio astronomy
Telstar
Radio astronomy
Taxi cab
Radio astronomy
10¹² Hz
10¹¹ Hz
10¹⁰ Hz
10⁹ Hz
10⁸ Hz
10⁷ Hz
Frequency

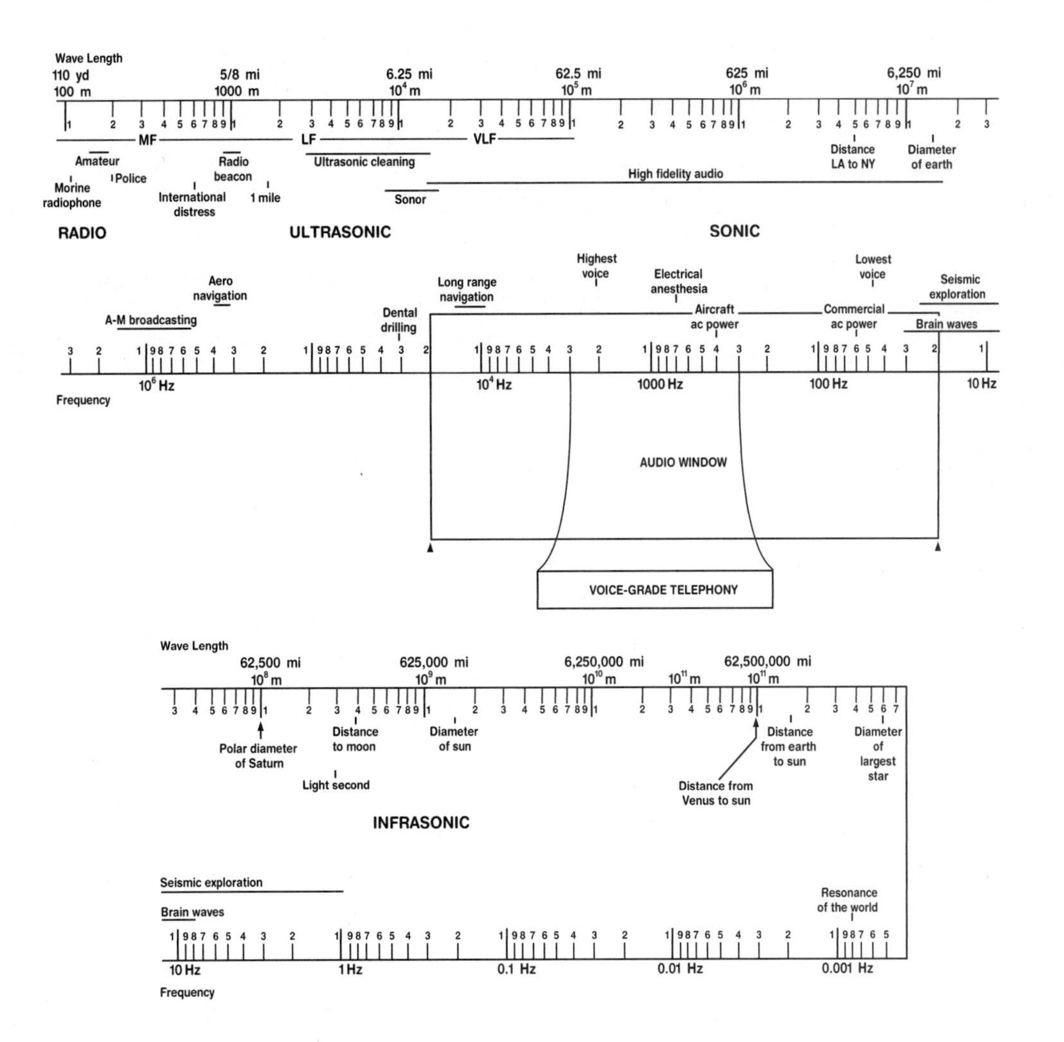
Wave Length
110 yd
100 m
5/8 mi
1000 m
6.25 mi
10⁴ m
62.5 mi
10⁵ m
625 mi
10⁶ m
6,250 mi
10⁷ m
MF
LF
VLF
Amateur
Morine radiophone
Police
Radio beacon
International distress
1 mile
Ultrasonic cleaning
Sonor
High fidelity audio
Distance LA to NY
Diameter of earth
RADIO
ULTRASONIC
SONIC
Aero navigation
A-M broadcasting
Dental drilling
Long range navigation
Highest voice
Electrical anesthesia
Aircraft ac power
Lowest voice
Commercial ac power
Seismic exploration
Brain waves
10⁶ Hz
10⁴ Hz
1000 Hz
100 Hz
10 Hz
Frequency
AUDIO WINDOW
VOICE-GRADE TELEPHONY
Wave Length
62,500 mi
10⁸ m
625,000 mi
10⁹ m
6,250,000 mi
10¹⁰ m
62,500,000 mi
10¹¹ m
Polar diameter of Saturn
Distance to moon
Light second
Diameter of sun
Distance from earth to sun
Distance from Venus to sun
Diameter of largest star
INFRASONIC
Seismic exploration
Brain waves
Resonance of the world
10 Hz
1 Hz
0.1 Hz
0.01 Hz
0.001 Hz
Frequency

IEEE LAN Committees

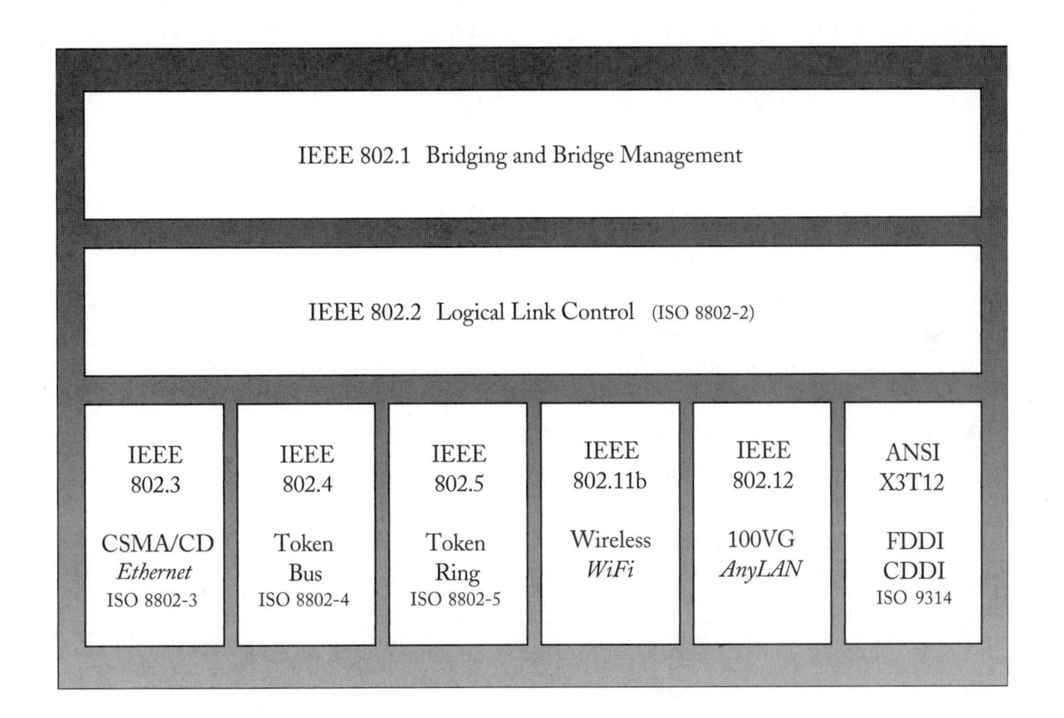

IEEE Project 802 Committees

802.1	Higher level Interfaces: Bridges & Management	
802.2	Logical Link Control (LLC)	
802.3	CSMA/CD *Ethernet* specifications	(Developed by DIX)
802.4	Token Passing Bus specifications	(Developed by GM)
802.5	Token Passing Ring specifications	(Developed by IBM)
802.6	Metropolitan Area Network specifications	[Incl. DQDB & SMDS]
802.7	Broadband Technical Advisory Group	[From 802.3b & 802.4]

802.8	Fibre Optic Technical Advisory Group	
802.9	Voice and Data Network specifications	[AKA. isoEthernet]
802.10	Security Encryption and Access Control	
802.11	Radio & Infrared Wireless LAN	[AKA. WiFi]
802.12	100VG *AnyLAN* (DPAM) specifications	(Developed by HP)
802.13	Skipped due to triskaidekaphobia	
802.14	CATV Broadcast Cable Modem	[AKA. DOCSIS]
802.15	Wireless Personal Area Network	[AKA. Bluetooth]
802.16	Broadband Wireless Access	[AKA. AirInterface]

IEEE LAN Committees (Institute of Electrical and Electronic Engineers)

A sampling of IEEE 802 Local Area Network Standards for Computer Interconnection.

802.1	Architecture & Overviews
802.1B	LAN/MAN Management
802.1D	Local MAC Bridging (Local—LAN to LAN)
802.1G	Remote MAC Bridging (Remote—LAN to LAN via WAN)
802.1H	MAC Bridging for Ethernet Version 2.0 in 802 LANs
802.1i	MAC Bridging for Fiber Distributed Data Interface (FDDI)
802.1j	Managed Objects for MAC Bridges (Supplement to 802.1D)
802.1k	Discovery and Dynamic Control of Event Forwarding (LAN & MAN)
802.1p	MAC Bridge supplement: Traffic Class Expediting, Dynamic Multicast Filtering
802.1Q	Virtual Bridged Local Area Networks (switched VLANs)
802.1q	Now 802.12e: An 802.1D supplement to support 802.12 100VG "AnyLAN"
802.2	Local Area Networks: Logical Link Control (ISO/IEC 8802-2)
802.2a	Flow Control Techniques for LAN Bridges
802.2b	Acknowledged Connectionless-mode Service Type 3 operation
802.2c	Logical Link Control Conformance Requirements LLC Type 1 & LLC Type 2
802.2f	LLC Sub-layer Management
802.2g	Supplement to 802.2, LLC Type 4 (High Speed, High Performance) operation

802.3	CSMA/CD LANs	(Ethernet, includes original 10BASE5 specification)
802.3a	10BASE2	Thin COAX (aka, ThinNet or CheaperNet)
802.3b	10BROAD36	Broadband COAX (obsolete)
802.3c	Repeater Unit	For 10 Mbps Baseband LANs
802.3d	FOIRL	Fiber Optic Inter Repeater Link (obsolete)
802.3e	1BASE5	Twisted Pair (StarLAN, also obsolete)
802.3i	10BASE-T	Unshielded Twisted Pair (UTP)

802.3j	10BASE-F	Fiber Optic Active and Passive Star-Based LANs (-FL, -FB, -FP)
802.3u	100BASE-T	Fast Ethernet —100BASE-TX, 100BASE-FX, and 100BASE-T4
802.3x	Supplement	Specification for 802.3 Full Duplex Flow Control
802.3y	100BASE-T2	Fast Ethernet —100 Mbps operation on 2 pairs of Cat3 UTP
802.3z	1000BASE-X	Gigabit Ethernet —1000BASE-SX, 1000BASE-LX, 1000BASE-CX
802.3aa	100BASE-T	Fast Ethernet Maintenance Revision
802.3ab	1000BASE-T	Gigabit Ethernet —1 Gbps operation over 4 pairs of Cat5 UTP

802.5	Token Ring Access Method and Physical Layer (4 Mbps/1985, 16 Mbps/1988)
802.5j	LAN: Fiber Optic Station Attachment
802.5m	Recommended practice Source Route Transparent bridging
802.5n	Standard for Operation of 4 and 16 Mbps Token Ring LANs on UTP media
802.5p	LAN: Part 2: Logical Link Control; End System Route Determination
802.5r	Revision of IEEE Standard 802.5 for dedicated, full duplex Token Ring operation

802.11	Wireless LAN/MAN — Provides 1, 2, 5.5, 11 and 54 Mbps wireless operation
802.11a	Wireless MAC & PHY specs: 54 Mbps in the 5 GHz band
802.11b	Wireless MAC & PHY specs: 1, 2, 5.5, 11 Mbps in the 2.4 GHz band (WiFi)
802.11c	MAC Bridges - Supplement for support by IEEE 802.11
802.11g	Wireless MAC & PHY specs: 1 - 54 Mbps in both the 2.4 and 5 GHz bands

APPENDIX C

Standard Data Rates

159.252 Gbps	OC-3072
39.813 Gbps	OC-768
13.271 Gbps	OC-256
9.953 Gbps	OC-192
4.976 Gbps	OC-96
2.488 Gbps	OC-48, STS-48
1.866 Gbps	OC-36
1.244 Gbps	OC-24
1 Gbps	IEEE 802.3 Gigabit Ethernet
933.12 Mbps	OC-18
622.08 Mbps	OC-12, STS-12
466.56 Mbps	OC-9
155.52 Mbps	OC-3, STS-3
100 Mbps	IEEE 802.3 Fast Ethernet, FDDI/CDDI, HSTRN
51.84 Mbps	OC-1, STS-1, (multiply by OC-# to get data rate)
44.736 Mbps	T-3, DS-3 North America
34.368 Mbps	E-3 Europe
16 Mbps	IEEE 802.5/Token Ring LANs
10Mbps	Classic Ethernet, 10Base5, 10Base2, 10Base-T
8.448 Mbps	E-2 Europe
6.312 Mbps	T-2, DS-2 North America
6.144 Mbps	Standard ADSL downstream
4 Mbps	IEEE 802.5/Token Ring LANs
3.152 Mbps	DS-1c
2.048 Mbps	E-1, DS-1 Europe

1.544 Mbps	T-1, DS-1 North America
128 kbps	ISDN
64 kbps	DS-0, pulse code modulation (PCM)
56 kbps	V.92 duplex modem
53 kbps	V.90, K56flex, U.S. Robotics x2 modems
33.6 kbps	V.34bis duplex modem
28.8 kbps	V.34, Rockwell V.Fast Class modems
19.2 kbps	V.32ter duplex modem
14.4 kbps	V.32bis duplex modem, V.17 fax
9600 bps	V.32 duplex modems
4800 bps	V.27ter duplex modem (2400/1200)
2400 bps	V.22bis duplex modem
1200 bps	V.22 duplex modem
600 bps	V.23 duplex modem (600/1200)
300 bps	V.21 duplex modem

ONLINE REFERENCES

For additional information, visit the following Web sites referenced in the text.

Web sites of general interest:

- www.BuddyShipley.com
- www.Anixter.com
- www.Cisco.com
- www.Siemon.com

Standards organizations' Web sites:

ANSI
www.ansi.org

ARIN
www.arin.net

EIA
www.eia.org

IEEE
www.ieee.org

IEC
www.iec.ch

ISO
www.iso.org

ITU
www.itu.int

ITU-T
www.itu.int/ITU-T

NEC
www.nfpa.org/nec

NEMA
www.nema.org

TIA
www.tiaonline.org

BIBLIOGRAPHY

International Standards documents:

ISO/IEC 8802-3 (ANSI/IEEE 802.3) CSMA/CD

ISO/IEC 10038 (ANSI/IEEE 802.1D) MAC Bridges

ISO 8802-2 (ANSI/IEEE 802.2) Logical Link Control

ISO/IEC 8802-5 (ANSI/IEEE 802.5) Token Ring Access Method

New York, New York: The Institute of Electrical and Electronic Engineers, 1985–2002.

Internetworking with TCP/IP, Volume 1, D. Comer.

Douglas E. Comer, Internetworking with TCP/IP, Volume 1, 3rd Edition. Englewood Cliffs, New Jersey: Prentice Hall, 1995.

Interconnections, R. Perlman.

Radia Perlman, Interconnections: Bridges and Routers, 2nd Edition. Reading, Massachusetts: Addison-Wesley, , 1992.

NetWare LAN Analysis, L. Chappell and D. Hakes.

Laura A. Chappell and Dan E. Hakes, NetWare LAN Analysis, 2nd Edition. Almeda, California: Sybex (Novell Press), 1994.

Novell's Guide to Troubleshooting NetWare 5, R. Sant'Angelo.

Rick Sant'Angelo, Novell's Guide to Troubleshooting NetWare 5. Indianapolis, Indiana: John Wiley & Sons, 1999.

Windows NT Server Survival Guide, R. Sant'Angelo.

Rick Sant'Angelo, Windows NT 4 Administrator's Survival Guide. Indianapolis, Indiana: Sams Publishing, 1997.

In addition, the following non-book sources of information were essential in authoring this text:

- The manuals, documentation, and Web sites produced by numerous manufacturers of network and internetworking products.
- More than 20 years of direct experience installing, maintaining, troubleshooting most of the technologies covered in this text.
- And over a decade of writing and teaching others about networks, protocols, and internetworking.

μ. Micrometer (0.000001 meters). A unit of measure commonly used to define the diameter of optical fiber. (Multi-mode fiber: 50μm–62.5μm; single-mode fiber: 8–10μm.)

AAL. ATM Adaptation Layer. Supports various levels of service: AAL Levels 1, 2, 3/4, and 5.

AD. Active Directory. Microsoft Windows directory services.

ADSL. Asymmetrical Digital Subscriber Line. Although developed to support multi-megabit data rates over a telephony infrastructure, most ADSL services provide downstream (receive) data rates from 256 kbps to 640 kbps and upstream (transmit) data rates from 128 kbps to 256 kbps. See DSL.

AMP. Active Monitor Present.

Amp. Ampere. A unit of measure for electrical current.

ANSI. American National Standards Institute (www.ansi.org). An independent standards body that usually works closely with IEEE.

AppleTalk. Apple Computer Corporation protocol suite. Initially developed to support Apple's proprietary LocalTalk LAN that uses CSMA/CA and operates at 230 kbps over Unshielded Twisted Pair (UTP). AppleTalk was later supported over other networks: Ethernet (EtherTalk), token ring (TokenTalk), and FDDITalk (FDDI).

ARP. Address Resolution Protocol. Developed by IETF, this protocol resolves 48-bit MAC addresses to 32-bit IP addresses. All IP hosts maintain an IP ARP cache.

ARPAnet. Advanced Research Projects Agency network. See DARPAnet.

AS. Autonomous System. Defines a single network controlled by a common set of routing protocols and policies.

ASCII. American Standard Code for Information Interchange. The predominant 7-bit (and 8-bit) data-encoding scheme used to provide binary representation of alphanumeric characters. See also EBCDIC and Unicode.

ASIC. Application Specific Integrated Circuit. Custom-developed electronic components that are engineered to perform a limited set of requirements. These highly specialized IC chips are commonly used in networking components (i.e., the PHY and MAC functions of an 802.3/Ethernet interface).

ASRT. Adaptive Source Route Transparent. A bridging protocol developed to provide connectivity between token ring LANs using Source Route Bridging and those using IEEE 802.1D transparent bridging.

ATM. Asynchronous Transfer Mode or cell relay. A high-speed network interface intended to satisfy virtually all current and future network communications requirements. The cost and complexity of ATM have limited its acceptance.

Attenuation. The loss of signal strength during transmission due to characteristics of the medium. Measured in decibels (dB).

AUI. Attachment Unit Interface, IEEE 802.3 (also called Ethernet Drop Cable). A Shielded Twisted Pair (STP) media using 15 pin "D" connectors, one male and the other female. The standard also specifies connectors that use a frequently unreliable slide latch mechanism to secure attachment of the cable to the transceiver (DCE) and host computer (DTE).

AWG. American Wire Gauge. A measurement of the diameter of a wire conductor. Higher gauge numbers indicate smaller wire diameters; lower gauge numbers indicate larger wire diameters.

Backbone. A context-sensitive term that, in a local context, is the LAN used to interconnect multiple workgroup LANs. From a WAN or enterprise-wide perspective, it may be a global data communications network that interconnects all remote sites. Also the top-level network in a hierarchically tiered system: a network of networks. Lower tiers of subordinate networks are attached to the backbone via bridges, switches, or routers.

Bandwidth. This term is commonly and mistakenly used to refer to the data rate of a given media. Technically, this term refers to *spectral bandwidth*, and is measured in cycles per second, or Hertz. In the context of this book, bandwidth defines the range of frequencies, or *spectrum*, used by a given network technology. The data encoding technique used to transmit data over this spectrum dictates the actual data-carrying capacity of the network (i.e., bits per second).

BEB. Binary Exponential Back-off algorithm. Part of the CSMA/CD protocol used by 802.3/Ethernet.

BECN. Backward Explicit Congestion Notification. Real-time flow control mechanism used by frame relay. See FECN.

BGP. Border Gateway Protocol. Used to provide internetwork routing between different Autonomous Systems, (e.g., BGP-4).

Blue Book. Name of the original documents defining the Digital, Intel, Xerox (DIX) specifications for Ethernet. (Bluebook 1 and Bluebook 2.)

BNC. Bayonet-Neill-Concelman (the names of the developers). Also Bayonet "N" Connector, or Baby "N" Connector, used in 10Base2 thin coaxial LANs.

BootP. Bootstrap Protocol. A TCP/IP network protocol and a mechanism used to dynamically configure IP host addressing. This protocol was replaced by DHCP.

bps. Bits per second. Bits are usually indicated by a lowercase "b" and bytes are indicated by an uppercase "B."

BRI. ISDN Basic Rate Interface. Supports 2 (64 kbps) "B" plus 1 (16 kbps) "D" channel over a 2-wire local loop telco circuit. See also ISDN and PRI.

BT. Bit Time. The transmission duration of a single binary value.

Burst. A function of some frame relay networks that allows the data transfer rate to exceed the CIR up to the specified port speed of the frame relay circuit.

c. Representation of the speed of light (186,282 miles/second or 299,792 km/second); for example, the nominal velocity of propagation for Cat5 UTP is .595c.

Capacitance. The capacity of a metallic wire or cable to store an electrical charge. Measured in farads (F) or picofarads (pF).

CAU. Control Access Unit. The central IBM token ring hub component that supports up to four token ring Lobe Access Modules (LAM).

CCITT. Comité Consultatif International Téléphonique et Télégraphique. This is now known as the ITU-T [for Telecommunication Standardization Sector of the International Telecommunications Union (www.itu.org)].

CD. Collision Detection portion of CSMA/CD in 802.3/Ethernet.

CD. Collision Domain. Includes all Ethernet LAN segments interconnected by repeaters or hubs.

CDDI. A technology for running 100 Mbps FDDI over copper using the ANSI X3T12 twisted pair physical medium dependent interface (TP-PMD).

CIDR. Classless Inter-Domain Routing. A newer method of IP addressing that avoids the restrictions of address classes to define networks and subnets.

CIR. Committed Information Rate. The rate of information transfer a subscriber has stipulated for a frame relay circuit. The CIR may be exceeded for short durations without risk of the network dropping frames. See also Burst.

CLNP. Connection-Less Network Protocol. The Network layer protocol developed by the ISO. Similar in function to IP, but supports 160-bit addresses (20 octets).

CMIP. Common Management Information Protocol (ISO 9595/9596). The network management protocol developed to support ISO OSI CMIS.

CMIS. Common Management Information Services. The ISO OSI system of network management proposed as a replacement for SNMP, which is used in IP-based internetworks. CMIS provides enhanced security and more extensible reporting capabilities, but it is not as widely accepted or supported as SNMP.

Collision. A normal event in CSMA/CD Ethernet that occurs when two or more devices transmit at the same time. All participating stations must detect collisions within the round-trip propagation delay time (51.2μs for 10 Mbps Ethernet). A late collision is one that has exceeded this round-trip time.

Concentrator. See Hub.

CRC. Cyclical Redundancy Check. Usually a 32-bit error detection algorithm used in LANs, such as Ethernet and token ring.

CSMA/CA. Carrier Sense, Multiple Access with Collision Avoidance. Used by Apple Computer Corporation's LocalTalk LAN.

CSMA/CD. Carrier Sense, Multiple Access with Collision Detection, (IEEE/ANSI 802.3, ISO/IEC 8802-3, and DIX ESPEC 1 and 2 Ethernet). Network access method or channel arbitration method (also known as the Ethernet protocol).

CSU. Channel Service Unit. A network component that replaces modems in digital WAN services. Sometimes referred to as an ISU (integrated CSU/DSU).

DAC. Dual Attached Concentrator. An FDDI interface specification for FDDI hubs.

DAS. Dual Attached Station. An FDDI interface specification for computers.

DARPAnet. Defense Advanced Research Projects Agency network. ARPAnet evolved into the global Internet.

dB. Abbreviation or notation for deciBel.

DCE. Data Communication End. Communicates to the Data Terminal End (DTE) of a circuit. Typically used for modems. See also DTE.

DDP. Datagram Delivery Protocol. AppleTalk routable Network layer protocol.

deciBel or decibel. A comparative (logarithmic) unit of measure of signal strength, usually the relation between a transmitted signal and a standard signal source.

DECnet. Digital Equipment Corporation's network protocol suite. Available in two versions: Phase/IV and Phase/OSI. DEC was acquired by Compaq/HP.

Demarcation. Generally used in reference to the point at which telecommunications and data communications (WAN) circuits are terminated. The physical location where building or equipment wiring interconnects with Local Exchange Carriers (LEC) and responsibility for service or maintenance transfers ownership.

DHCP. Dynamic Host Configuration Protocol. An extension of BootP that offers dynamic allocation of IP addresses, IP addressing details, and related information. DHCP provides safe, reliable, and simple TCP/IP network configuration, prevents address conflicts, and helps conserve the use of IP addresses through centralized management of address allocation.

DIW. Type D (or Data) Inside Wire. Originated as specific AT&T telephony medium, now commonly used to describe any 2, 3, or 4 pair UTP cable where the copper conductors are 22, 24, or 26 AWG with a PVC jacket.

DIX. Abbreviation for Digital, Intel, Xerox which were the original developers of ESPEC1 and ESPEC2 Ethernet.

DLCI. Data Link Connection Identifier. The 2-byte frame relay circuit connection address field.

DLS/R. Dynamic LU Server/Requester. Used in IBM SNA environments.

DLSw. Data Link Switching. Unrelated to the OSI Data Link layer or switching, it is used in IBM environments to encapsulate the non-routable SNA protocol inside TCP/IP so that it can be routed; also provides local LLC termination to spoof session polling.

DNA. Digital Network Architecture. See DECnet.

DNS. Domain Name System or Service. Sometimes referred to as the BIND service in Unix systems. DNS provides a hierarchical name service for TCP/IP hosts. The DNS is configured with a list of host names and IP addresses, allowing IP hosts to query the DNS to specify remote systems by hostnames rather than IP addresses. Although DNS is also supported by Microsoft, IP domains are not to be confused with Microsoft Network Domains (i.e., Active Directory).

DoD. Department of Defense, USA. See TCP/IP and DARPAnet.

DPAM. Demand Priority Access Method. Devised by HP for 802.12 100VG AnyLAN, supports two access priority levels, normal and high.

DQDB. Distributed Queue Dual Bus. Defined for use with IEEE 802.6 MAN.

Drop Cable. Cable used to connect Ethernet stations or DTEs to a transceiver. See AUI.

DRP. DECnet Routing Protocol. The DECnet Phase/IV distance-vector routing information protocol.

DS0. Digital Service Level 0 (64 kbps).

DS1. Digital Service Level 1 (1.544 Mbps).

DS3. Digital Service Level 3 (44.736 Mbps).

DSL. Digital Subscriber Line. A local loop technology used to provide broadband Internet access now offered by many telcos and ISPs. Also known as HDSL (for High-speed DSL); IDSL (for DSL over ISDN); or xDSL (where "x" is the variant du jour). See also ADSL and SDSL.

DSS. Domain SAP Server. Used in Novell NetWare/IP environments, it provides the SAP function normally associated with IPX.

DSU. Digital Service Unit. DCE that replaces modems in digital WAN services. Often referred to as CSU/DSU. Also known as ISU and integrated CSU/DSU.

DTE. Data Terminal End. Communicates to the Data Communication End (DCE) of a circuit. Typically used for computers and terminals. See also DCE.

EBCDIC. Extended Binary Coded Decimal Interchange Code. IBM's 8-bit data-encoding scheme used to provide binary representation of alphanumeric characters. See also ASCII and Unicode.

EGP. Exterior Gateway Protocol. The actual name of an older EGP protocol used to provide internetwork routing between Autonomous Systems. See also BGP.

EGP. Exterior Gateway Protocol. Generic term used to define internetwork routing protocols used between autonomous systems (in contrast to IGP).

EIA. Electronics Industry Association. See EIA/TIA 568-A.

EIA/TIA 568-A. Commercial Building Telecommunications Cabling Specifications. Includes Category 3, 4, 5, 5E media and component specifications, structural definitions, installation requirements, and certification criteria.

EIGRP. Enhanced Interior Gateway Routing Protocol. A routing information protocol developed by Cisco Systems as an improvement of IGRP.

ELFEXT. Equal Level Far End Cross Talk. A calculated result (not a measurement). It is derived by subtracting the attenuation of the disturbing pair from the Far End Crosstalk (FEXT) that a given pair induces into an adjacent pair. Because 1000Base-T Gigabit Ethernet uses all four pairs of a Cat5/5E cable, ELFEXT was specified to quantify these effects.

EMI. Electro-Magetic Interference. See also RFI.

ES-IS. End System to Intermediate System. Developed by the ISO to work in conjunction with IS-IS as part of the OSI Network layer protocol, CLNP. ES-IS provides end stations with host address information similar to how ARP works with IP.

ESPEC1. The first DIX Ethernet specification (Bluebook 1).

ESPEC2. The second DIX Ethernet specification (Bluebook 2, and 802.3 draft proposal).

Ethernet. The name commonly used to refer to the ISO/IEC 8802-3 ANSI/IEEE 802.3 standard and CSMA/CD LANs (DIX ESPEC). The Xerox Corporation owns the name.

Extranet. Special networks usually configured to support Web-based Electronic Data Interchange (EDI) between different companies' internetworks.

Fast Ethernet. IEEE 802.3u 100 Mbps implementation of Ethernet. Supports 100Base-FX, 100Base-FX, 100Base-T4, and 100Base-T2. TX and FX are most common.

FCS. Frame Check Sequence. A field in all LAN frames that contains the CRC value of the rest of the fields in a frame.

FDDI. Fiber Distributed Data Interface. B203 standard that defines a 100 Mbps, dual counter-rotating, fiber optic, token ring passing LAN. Supports optical fiber, STP and UTP. See CDDI.

FECN. Forward Explicit Congestion Notification. Real-time flow control mechanism used by frame relay. See BECN.

FEXT. Far End Cross Talk. Signal distortion caused by the coupling of an outgoing signal at the receiving end of a circuit with the incoming signal being transmitted from the other end of the circuit. See NEXT, MD-NEXT, PS-NEXT, ELFEXT.

FIFO. First In, First Out. The buffer method used in Gigabit Ethernet buffered distributors.

FLP. Fast Link Pulse. An automatic test conducted by 802.3 100Base-T transceivers to determine link speed (10 or 100 Mbps), to determine if full or half-duplex operation can be supported, and to monitor the Link Segment.

FOIRL. Fibre Optic Inter-Repeater Link. IEEE 802.3 inter-repeater fiber specification. (Superceded by 10Base-FL.)

Fragment. An Ethernet transmission of between 96 and 511 bits. Shorter than the minimum frame size of 512 bits (or 64 bytes). Always interpreted as the result of a collision (also known as packet fragment).

Frame. The MAC sublayer portion of a local area network (LAN) protocol data unit (PDU). This term is frequently used interchangeably, however inaccurately, with the term "packet." Frames include a protocol header and an upper layer data payload, followed by a protocol trailer. Packets lack the trailer. See packet.

FQDN. Fully Qualified Domain Name. The complete alphanumeric identifier of a specific IP host, for example, www.buddyshipley.com. "www" identifies a specific IP host; "buddyshipley" identifies a specific Internet domain; and ".com" identifies the top level domain in which "buddyshipley" is included.

Frame Relay. A high-speed switched wide area network service. (1) A 1.5 layer protocol similar to X.25 that provides access speeds up to T-1 and T-3. Frame relay supports PVC, SVC, DLCI, FECN, BECN, Port Speed, CIR, and Burst. Many frame relay providers do not support all of these features. As frame relay networks have become more congested, the benefit of the Burst feature is rarely realized, therefore, rendering CIR is useless. Many frame relay circuits are now defined only by DLCI and Port Speed. Frame relay is defined as a connection-oriented Layer 2 protocol specified by the ITU-T for the transfer of information between two compatible endpoints. The frame relay protocol is referred to as LAPF, which is a simplified subset of LAPD (also based on HDLC), but defines a simplified protocol with frame delimiters or flags, virtual Data link Connection Identifiers (DLCIs), congestion indication, discard eligibility bits, and error detection capability. The variable length frame can typically be up to 4096 bytes long. Frame relay does not include control procedures, such as retransmission or flow control, and is optimized for low error-rate networks. Frame relay is specified in ITU-T Recommendation Q.922. PVC management was first specified in LMI, then ANSI T1.617, and finally Recommendation Q.922. (2) The name used to describe different carriers' data transport services that are based on the use of the frame relay protocol. Customers typically subscribe to frame relay to replace or extend the use of private line data networks. Since frame relay specifies a protocol, not a service, offerings from the various carriers may vary.

FTP. Foil Twisted Pair. Copper cabling similar to ScTP or STP; it uses a thin foil shield rather than a heavy gauge braided metal.

FTP. File Transfer Protocol. A protocol and utility used to move and manage files between IP hosts. Considered part of the TCP/IP suite of protocols.

Gbps. Giga (billion) bits per second. Also Gb/s.

Gig-E. Gigabit Ethernet. IEEE 802.3z 1000 Mbps implementation of Ethernet, supports 1000Base-SX, 1000Base-LX, 1000Base-CX, and 1000Base-T. T and SX are the most common.

GMII. Gigabit Medium Independent Interface. Gigabit Ethernet equivalent to the 100 Mbps Fast Ethernet MII cable/interface. A Gigabit Ethernet network interface adapter may support the GMII in addition to an embedded transceiver. The GMII interface may support any of the following external transceivers: 1000Base-SX, 1000Base-LX, 1000Base-CX, and 1000Base-T.

HDLC. High-level Data Link Control. International standard bit-oriented WAN communications protocol developed by the ISO, used in one form or another by internetwork device manufacturers. HDLC is the default frame format used by Cisco Systems routers for DS-0 and DS-1 (T-1) interfaces.

HTTP. Hyper Text Transfer Protocol. The primary IP protocol used to support World Wide Web applications. Web browsers use HTTP to access Web servers.

Hub. Usually defines a multi-port repeater, such as "a 10Base-T hub." Sometimes referred to as a concentrator.

IAB. Internet Architecture Board. Oversees Internet standards and protocol development efforts of the IETF and IESG.

IDF. Intermediate Distribution Frame. Another term for telecommunications closets distributed throughout a building.

IEC. International Engineering Consortium.

IEC. Inter-exchange Carrier. A common carrier for international communications. See also IXC.

IEEE. Institute of Electrical and Electronic Engineers. An independent standards body that defined 802.2 Logical Link Control, 802.3/Ethernet, 802.4 token bus, 802.5 token ring, 802.12 100VG AnyLAN standards, and others.

IESG. Internet Engineering Steering Group. Defines the strategic direction for Internet protocol standards. See Internet Architecture Board (IAB).

IETF. Internet Engineering Task Force. Develops the actual Internet protocol standards. See Internet Architecture Board (IAB).

IFG. Inter-Frame Gap. A pause between 802.3/Ethernet frames (9.6μs for 10 Mbps).

IGP. Interior Gateway Protocol. Routing information protocols used within an Autonomous System, such as RIP or OSPF.

IGRP. Interior Gateway Routing Protocol. Routing information protocol developed by Cisco Systems as an improved alternative to standard RIP.

Impedance. The total opposition a circuit, cable, or component offers to alternating flow. It includes both resistance and reactance and is generally expressed in ohms.

Internet. The global TCP/IP-based internetwork that evolved from DARPAnet in the late 1960s and today supports the World Wide Web, telephony, and other business and recreational applications too numerous to mention.

Internetworking. Encompasses all of the services required to provide connectivity and interoperability between network systems. The term "internetworking" encompasses all required levels of communications between similar and dissimilar network systems; therefore, there is no single, concise definition.

Interoperability. Provides users with platform independent access to all applications and resources regardless of which mix of systems they reside on. Resources provided on different computing platforms are transparent to the user's native operating environment. These systems may be connected to the same network or via a private internetwork or intranet, or the public Internet.

Intranet. Private networks that employ TCP/IP and World Wide Web protocols and other Internet-based technologies.

IP. Internet Protocol. The routable Network layer protocol developed by the IETF as the foundation of Internet communications.

IPX. Internetwork Packet eXchange. The network protocol developed by Novell, Inc. for use with the company's NetWare network operating system.

ISDN. Integrated Services Digital Network. Supports broadband, primary rate, and basic rate interfaces (PRI & BRI). PRI uses the 24 DS-0 channels of a 4-wire T-

1 circuit to support 23 "B" channels and 1 "D" channel (64 kbps each channel). BRI uses a 2-wire circuit to support 2 "B" channels (64 kbps) plus 1 "D" channel (16 kbps). The two "B" channels may be bonded to provide an aggregate of 128 kbps. Bonding all three channels provides 144 kbps to support IDSL (ISDN DSL).

IS-IS. Intermediate System to Intermediate System. A link state routing information protocol developed by the ISO. IS-IS is a similar to OSPF and works in conjunction with ES-IS as part of the OSI network layer protocol, CLNP.

ISO. International Standards Organization (or: International Organization for Standardisation). An independent standards body that defined the OSI seven layer reference model and related suite of protocols.

ISP. Internet Service Provider. Sells Internet access to the general public through a variety of means. ISP customers may use common analog dial-up modems to access the ISP's network, or in many areas, subscribers may opt for high-speed broadband access such as DSL or cable modem. If these services are unavailable, ISDN or wireless solutions are fair alternatives.

ITU. International Telephony Union (or ITU-T), formerly the CCITT. Responsible for the X.nn and V.nn standards (e.g., X.25 and V.34). Many ITU/CCITT standards have been adopted by the ISO for use with OSI.

IXC. Inter-eXchange Carrier. A common carrier for international communications. See also IEC.

kbps. Expressed in lowercase "k", kilo (thousand) bits per second, also kb/s. The precise meaning of "kilo-" depends on the context in which the term is used. Outside of computers and data communications, one kilo always means a unit of 1000. In the world of computers however, the term kilo may sometimes refer to a unit of 1000, or a binary multiple of 1024 (= 2^{10}). Kbps is commonly found in reference to memory and hard drive capacities.

Kludge. A system, especially a computer system that is constituted of poorly matched elements or of elements originally intended for other applications. This term, found frequently in the jargon of the engineering and computer professions, denotes a usually workable but makeshift system, modification, solution, or repair. Generally considered undesirable; avoid when at all possible.

LAM. Lobe Access Module. The IBM token ring hub component that supports up to 20 token ring Lobe (station) interfaces each.

LAN. Local Area Network. Usually supports a small geographic area (<2500m), high speed (4–1000 Mbps), and the user owns the media. Includes Ethernet, Fast Ethernet, Gigabit Ethernet, token ring, FDDI/CDDI, WiFi, and ATM.

LAPB. Link Access Procedure Balanced. WAN frame format prescribed by the ISO.

LAPD. Link Access Procedure-D.

LASER. Light Amplification by Stimulated Emission of Radiation. The light source used in many multi-mode and all single-mode fiber transmitters. See also LED.

LAT. Local Area Transport. A non-routable DECnet/IV protocol that supports terminal server to host communications.

LATA. Local Access and Transport Area. The territory serviced by a given local telephone company.

LEC. Local Exchange Carrier. A local telephone company. Some LECs are regional Bell operating companies (RBOCs).

LED. Light Emitting Diode. The light source used in most mutli-mode fiber transmitters. See also LASER.

Link Segment. A length of UTP or optical fiber media. Used in point-to-point Ethernet connections (one device at each end).

Link Status. An automatic test conducted by 802.3 10Base-T transceivers to monitor the condition of the Link Segment.

LLC. Logical Link Control. The top half of OSI Data Link layer (layer 2). This protocol sublayer may be used with all MAC sublayer protocols such as 802.3/Ethernet, 802.5 token ring, or X3T125 FDDI/CDDI.

Local Loop. In telephony, the Twisted Pair Copper circuit connecting the subscriber to the Local Exchange Carrier's Central Office (CO).

LocalTalk. Apple Computer Corporation's proprietary CSMA/CA-based LAN protocol that operates at a screaming 230 kbps.

MAC. Media Access Control. Bottom half of OSI Data Link layer (layer 2).

MAC. Moves, Adds and Changes. Refers to cable plant modifications.

Mac. Nickname for Apple Corporation Macintosh personal computers.

MAN. Metropolitan Area Network. Supports larger geographic areas than a LAN (~50 km), moderate to high speeds (56 kbps–155 Mbps and higher), and LEC owns the media. Technologies include SONET, ISDN, 802.6 and ATM.

MAU. Medium Attachment Unit. IEEE 802.3 standards term for the Ethernet transceiver component/function.

MAU. Multi-station Access Unit. IEEE 802.5 standards term for token ring wiring hub (MSAU).

Mbps. Expressed in uppercase "M", Mega (million) bits per second, also Mb/s.

MDF. Main Distribution Frame. Another term for the primary telecommunications equipment room in a building.

MDI. Medium Dependent Interface. 802.3 Ethernet interface specification that defines the transceiver-to-media physical coupling. In twisted pair Ethernet, MDI refers to a straight-through pinout, whereas MDI-X refers to a crossover function.

MD-NEXT. Multiple Disturber NEXT. Distortion between adjacent pairs in a common cable. See also NEXT and PS-NEXT.

MHz. Megahertz. Unit of frequency equal to one million cycles per second, a measure of spectral bandwidth.

MIB. Management Information Base. Includes MIB-I, MIB-II, RMON, and proprietary MIB extensions. Used with (SNMP).

MIC. Medium Interface Connector. IEEE 802.5 standards term for the token ring shielded twisted pair cable jack and plug known as the Universal Data Connector.

MII. Medium Independent Interface. 802.3 100Base-T Fast Ethernet equivalent to the 10 Mbps AUI cable and interface. A Fast Ethernet network interface adapter can support the MII in addition to an embedded transceiver. The MII interface can support any of the following external transceivers: 100Base-TX, 100Base-T4, and 100Base-FX.

MMF. Multi-Mode Fiber. Optical fiber media, typically: 50/125μm–62.5/125μm

NBP. Name Binding Protocol. Announces and gathers NetBIOS Names of each machine in a NetBIOS network.

NCP. NetWare Core Protocol. Provides access to shared file and print resources in Novell NetWare networks.

NCP. Network Control Program. The operating system used in IBM Front End Processors (FEP).

NetBEUI. NetBIOS Extended User Interface. Microsoft's term for the native NetBIOS LAN interface and protocol.

NetBIOS. Network Basic Input Output System. A non-routable protocol, developed by Sytek for IBM to use in its early PC LAN network. Requires each machine to have a unique 15-character NetBIOS name and supports SMB and NBP.

NEXT. Near End Cross Talk. Signal distortion caused by the coupling of an outgoing signal at the originating end of a circuit with the incoming signal being received from the other end of the circuit. See also PS-NEXT and MD-NEXT.

NIC. Network Interface Card. Usually includes an embedded transceiver for a specific type of media. Referred to as a Network Controller Card in larger computer systems and often does not include an embedded transceiver (i.e., AUI interface only).

NLSP. NetWare Link Services Protocol. A Link State routing protocol developed by Novell for use with IPX. NLSP is a similar to OSPF and IS-IS and incorporates the SAP function as well as Link State routing.

nm. Nanometer (0.000000001 meters). Unit of measure commonly used to define the velocity of signal propagation across a given media.

NOS. Network Operating System. Like an operating system (OS) but with additional network-specific server capabilities and features. It is the program that manages all of the other application programs in the server, provides access to that computer's resources and peripherals, and manages network attached resources. Common NOSs include: Novell, NetWare, Banyan Systems VINES, Microsoft Windows NT/2000/2003 Servers.

NVP. Nominal Velocity of Propagation, Velocity of Propagation, or Propagation Delay. The speed at which a signal travels through a medium relative to the speed of light, "c." Expressed as a percentage of the speed of light where c = approximately 186,000 miles per second (300,000 kilometers per second).

OS. Operating System. The program that manages all of the other application programs in a computer and provides access to that computer's resources and peripherals. Common OSs include Microsoft's DOS, Windows 95/98/Me, Windows NT/2000/XP, Apple's Mac OS, all of the varieties of Unix and Linux.

OSI. Open Systems Interconnection. The ISO's approach to a layered network protocol architecture. Refers to the OSI seven layer reference model and the ISO's suite of network protocols.

OSPF. Open Shortest Path First. A Link State routing information protocol developed by the IETF primarily for use with IP.

OUI. Organizationally Unique Identifier. Defined by the IEEE as the first three bytes of the 6-byte MAC address. Used to define the vendor or manufacturer of a network device.

Packet. The term is most commonly used to refer to network layer PDUs (OSI layer 3), for example "an IP packet." The term "packet" is frequently used interchangeably (albeit incorrectly) with the term "frame." See frame.

PARC. Xerox Palo Alto Research Center (Xerox PARC). Birthplace of Ethernet.

PDA. Personal Digital Assistant. A handheld computer used to manage schedules, contacts, and tasks. Also known as a Personal Information Manager (PIM).

PDS. Premises Distribution System. A structured cabling system developed by AT&T.

PDS 110. The AT&T PDS cabling system components, punchdown blocks, and connectors.

PDU. Protocol Data Unit. A protocol header and upper layer data payload. For example, an IP PDU may also contain TCP and application data.

PHY. Refers to Physical layer network components as defined by the ISO OSI. The Physical layer chipsets developed to support the 10/100/1000 Mbps 802.3/Ethernet protocols are referred to as "PHY chips."

PIM. Personal Information Manager. A handheld computer used to manage schedules, contacts, and tasks. Also known as a Personal Digital Assistant (PDA).

Plenum Cable. Plenum rated cabling. An alternative material used as cable jackets or sheathing and as insulation. Considered safer than PVC cable because it does not release toxic fumes when burned. Initially designed for installation in air return (plenum) spaces above ceilings and below floors. Although it tends to be more expensive and difficult to work with, most electrical building codes require the use of plenum-rated cabling rather than PVC.

PPP. Point to Point Protocol. Used primarily for Internet dial-up in place of SLIP designed to provide multi-vendor internetwork interoperability, to support dynamic IP address allocation and to provide authentication security.

PRI. ISDN Primary Rate Interface. Supports 23 (64 kbps) "B" channels plus 1 (64 kbps) "D" channel over a 4-wire DS-1 (T-1) telco circuit. See also ISDN and BRI.

PSELFEXT. Power Sum ELFEXT is another calculation (not a measurement). PSELFEXT is derived from an algebraic summation of the individual ELFEXT effects on each pair by the other three pairs, and results in four separate PSELFEXT results for each end.

PS-NEXT. Power Sum NEXT. The cumulative distortion caused when all pairs in a common cable are energized simultaneously. See NEXT, MD-NEXT, ELFEXT.

PVC. Permanent Virtual Circuit. A static logical path through switched networks such as ATM, X.25, or frame relay.

PVC. Polyvinyl Chloride. A common material used as cabling jackets and insulation. Considered hazardous because it releases toxic fumes when burned. Most electrical building codes require the use of plenum-rated cabling.

QoS. Quality of Service. A feature of traffic flow control systems intended to control finite network bandwidth to provide delay-sensitive applications with higher network access priority.

RARP. Reverse Address Resolution Protocol. Resolves 32-bit IP addresses to 48-bit MAC addresses. Developed by IETF.

RBOC. Regional Bell Operating Company. Descendents of AT&T after divestiture which include: Ameritech, Bell Atlantic, Bell South, NYNEX, Pacific Bell, South Western Bell (SBC), and Qwest. All RBOCs are LECs.

RD. Route Discovery. A function in 802.5/token ring to identify a path across multiple token ring LANs interconnected by Source Route Bridges.

RFC. Request for Comment. Draft specifications and standards of the Internet Engineering Task Force (IETF).

RFI. Radio Frequency Interference. See also EMI.

RG. Radio Guide, as in RG-58a/u, RG-58c/u, and RG-59 coaxial cabling.

RIP. Routing Information Protocol. A distance-vector routing information protocol (or interior gateway protocol) supported in one form or another by many network protocols such as IP, IPX, and XNS. IP supports both RIP-1 and RIP-2.

RTMP. Routing Table Maintenance Protocol. AppleTalk distance-vector routing information protocol.

Runt. A packet fragment that is less than 96 bits. (Various vendors measure and define these events differently.)

SAA. Systems Application Architecture (son of SNA). Includes support for LU6.2 and APPC/APPN.

SAP. Service Advertising Protocol. Announces the availability of services to all stations of a Novell NetWare network.

SAS. Single Attached Station. An FDDI interface specification for computers.

ScTP. Screened Twisted Pair. Copper cabling similar to STP; usually has a lighter gauge foil shielding and is sometimes referred to as Foil Twisted Pair (FTP).

SDLC. Synchronous Data Link Control. Bit-oriented WAN communications protocol used in IBM SNA environments.

SDSL. Single-line DSL (original usage), or Synchronous DSL.

Segment. Bus segment. A length of media (usually coax), terminated at each end, that can support many devices connected via taps along the length of the media. A segment can consist of multiple sections of media. See also Link Segment.

SFD. Start of Frame Delimiter. LAN frame demarcation field between frame timing header and the MAC PDU.

SLIP. Serial Line IP. At one time, commonly used for Internet dial-up and requires defining a static IP address for all potential hosts; superceded by PPP.

SMB. Server Message Block. Provides access to shared file and print resources in NetBIOS networks.

SMDS. Switched Multi-megabit Data Service. Used with IEEE 802.6 MAN.

SMF. Single-Mode Fiber. Optical fiber media, typically smaller than MMF: 8–10 µm

SNA. Systems Network Architecture. Refers to all of IBM Corp. computer communications technologies, and IBM's proprietary network protocol suite.

SMP. Standby Monitor Present. A function in token ring LANs used to identify the presence of all Standby Monitor stations.

SMT. Station Management specified in the ANSI X3T12 FDDI/CDDI protocols.

SNMP. Simple Network Management Protocol. The de facto standard, vendor-independent management protocol supported by virtually all network vendors. Originated for use with TCP/IP networks, it has proliferated into other networking protocols such as IPX.

SOHO. Small Office, Home Office. Generally refers to networks and networking equipment intended to support very few users and computers.

SPX. Sequenced Packet Exchange. Augments the capabilities of Novell's IPX protocol; also referred to as IPX/SPX.

SRB. Source Route Bridge. A bridging option designed to support only 802.5/token ring.

Standard. (de facto) Specifications that have achieved such recognition based on their own merits that a majority of independent vendors have chosen to support and service them, not officially sanctioned by any standards body.

Standard. (de jure) Industry or formal specifications that have been developed, refined, and ratified with a majority vote by an independent group such as the IEEE, ANSI, ITU, or ISO. Members of these groups consist of manufacturers, end users, and other parties having a vested interest.

STP. Spanning Tree Protocol. Part of IEEE 802.1D MAC bridge specifications.

STP. Shielded Twisted Pair. Used as a LAN medium in some IBM and other token ring installations.

STP-A. An enhanced variety of Shielded Twisted Pair engineered to support higher transmission rates; used as a LAN medium in some IBM and other token ring installations.

SVC. Switched Virtual Circuit. A dynamic logical path through switched networks such as ATM, X.25, or frame relay.

TCP/IP. Transmission Control Protocol/Internet Protocol. The network communications protocol defined by the IETF and used by the Internet. Today, TCP/IP support is integrated into virtually all computer operating systems. It is the de facto standard protocol suite in use today and provides interoperability in multi-vendor computing environments and is now supported in some PDA/cell phones and other personal information managers (PIM).

TIA. Telecommunications Industry Association. See EIA/TIA 568-A.

TLA. Three Letter Acronym or Tiny Little Abbreviation (the vast majority of terms in this glossary).

TP-PMD. Twisted Pair - Physical Medium Dependent interface. Part of the ANSI X3T12 CDDI specifications to support FDDI over twisted pair copper media.

Trunk. Refers to thick coaxial cable (Linear Bus coax). Defined by IEEE 802.3 10Base5 Ethernet.

TSB. Technical Service Bulletin. Supplemental specifications to draft standards of ANSI/EIA/TIA.

Unicode. An alternative 16-bit data-encoding scheme used to provide binary representation of very large character sets. See also EBCDIC and Unicode.

UNC. Universal Naming Convention. Used in most microcomputer LANs to identify shared network resources such as file and printer servers, for example, \\ServerName\ShareName.

USOC. Uniform Service Order Code, the original telco cabling pin-out standard; obsolete.

UTP. Unshielded Twisted Pair. Copper cable available in voice and data grades, used as LAN and telephony medium. Contemporary UTP cabling adheres to electrical specifications defined by the EIA/TIA standards such as Categories 3 and 5E.

VG. Voice Grade. Refers to UTP cabling (also, 100VG-AnyLAN).

VIP. Vines Internet Protocol. The network protocol developed by Banyan Systems for use with the Vines NOS.

VLAN. Virtual LAN. Technologies and methods of managing multiple logical networks on one common physical network. Virtual LANs are isolated from each other, although they are actually sharing the same physical LAN switches and cable plant.

VRTP. Vines Routing Table Protocol. The Banyan Vines distance-vector routing information protocol.

WAN. Wide Area Network. Supports extended geographic area (global), moderate to high speeds (9.6 kbps to 56 kbps to 1.544 Mbps and higher). IXC/common carrier owns the media. Includes: Private Line, X.25, and Frame Relay.

WiFi. Wireless Fidelity. A marketing term used to identify IEEE 802.11b-compatible wireless LAN technologies.

WINS. Windows Internet Name Service. A name resolution service used in Microsoft Networking environments to resolve NetBIOS names to IP addresses in a routed environment. WINS is like a DNS for NetBIOS names.

WinSock. Windows Sockets. Developed by Microsoft, WinSOCK is the adaptation of the Berkeley Unix sockets application programming interface (API) to the family of Microsoft Windows operating systems. Sockets are a popular programming convention for communicating between applications across TCP/IP networks.

XNS. Xerox Network System. The network protocol developed by Xerox Corporation; also supports RIP.

INDEX

G

H

I

O

P

Q

R

U

V

W

X

Z